Advances in Science, Technology & Innovation

IEREK Interdisciplinary Series for Sustainable Development

Advances in Science, Technology & Innovation (ASTI) is a series of peer-reviewed books based on the best studies on emerging research that redefines existing disciplinary boundaries in science, technology and innovation (STI) in order to develop integrated concepts for sustainable development. The series is mainly based on the best research papers from various IEREK and other international conferences, and is intended to promote the creation and development of viable solutions for a sustainable future and a positive societal transformation with the help of integrated and innovative science-based approaches. Offering interdisciplinary coverage, the series presents innovative approaches and highlights how they can best support both the economic and sustainable development for the welfare of all societies. In particular, the series includes conceptual and empirical contributions from different interrelated fields of science, technology and innovation that focus on providing practical solutions to ensure food, water and energy security. It also presents new case studies offering concrete examples of how to resolve sustainable urbanization and environmental issues. The series is addressed to professionals in research and teaching, consultancies and industry, and government and international organizations. Published in collaboration with IEREK, the ASTI series will acquaint readers with essential new studies in STI for sustainable development.

More information about this series at http://www.springer.com/series/15883

Abdul Hai Alami

Mechanical Energy Storage for Renewable and Sustainable Energy Resources

 Springer

Abdul Hai Alami
Sustainable and Renewable Energy Engineering
University of Sharjah
Sharjah, United Arab Emirates

ISSN 2522-8714   ISSN 2522-8722   (electronic)
Advances in Science, Technology & Innovation
IEREK Interdisciplinary Series for Sustainable Development
ISBN 978-3-030-33790-2   ISBN 978-3-030-33788-9   (eBook)
https://doi.org/10.1007/978-3-030-33788-9

This Springer imprint is published by the registered company Springer Nature Switzerland AG
The registered company address is: Gewerbestrasse 11, 6330 Cham, Switzerland

To Seif. Hoping the world is a much better place when you're in the driver seat.

Foreword

As a former director and founding member of the Institute of Engineering and Energy Technologies (www.uws.ac.ukieet) at the University of the West of Scotland, I find myself very enthusiastic about the vantage point of this book. Mechanical energy storage for renewable resources is an important subdivision of the vast field of energy storage in general. Ever since I founded of the International Conference on Sustainable Energy and Environmental Protection (SEEP, www.seepconference.co.uk), energy storage has been the pinnacle of many important submissions and roundtable discussions on the subject of energy conversion. This is also true as I have the privilege of being a Subject Editor of the Elsevier Energy Journal and being a board member of a few other journals, where many papers that have been published and also those who are being submitted to the journal prove the importance of the storage topic. I can attest that having a book that is uniquely dedicated to mechanical storage is long overdue. The book chapters take the reader through a journey that makes all the stops, starting from energy harvesting right through utilization. The book emphasizes the need to think of storage technologies outside the conventional confines of battery storage. Granted, the portability and unanimous global adoption of batteries is the de facto storage technique for a world that converses almost exclusively through electrical generation and consumption, but other storage technologies are currently rising to prominence and deserve to be candidates for a future that has many challenges when it comes to energy. From compressed-air energy storage (CAES) systems to mechanical flywheels, the book lays an excellent mathematical foundation that is appealing to both engineering students and energy storage professionals alike. Some new technologies are also presented that are both thought-provoking and also promising, like buoyancy work energy storage that, if applied on a large scale, is destined to be very competitive especially for offshore maritime wind farms.

This book is an important addition to an energy engineer library. As a chairman of the Sustainable and Renewable Energy Engineering Department at the University of Sharjah, our students themselves are hybrid of electrical and mechanical engineers. I am certain that they will benefit from this book during their time at the department and beyond, and I am sure that the message the book carries would also appeal to the international audience as well. I have known Abdul Hai as an active member of the department who enjoys his teaching duties as much as he enjoys doing his research with his students. It is clear that he had them in mind while putting together the scientific matter in this book, and I am confident that the reader will find this clear throughout the chapter and pages of this book.

Abdul Ghani Olabi Ph.D.
Chairman of the Sustainable and Renewable Energy Engineering Department
University of Sharjah
Sharjah, United Arab Emirates
e-mail: aolabi@sharjah.ac.ae

Preface

Energy and water resources are the two most significant topics that concern humanity as a whole. Although their importance has been highlighted and associated with wars and power struggles throughout history, their sufficiency and sustainability are focal points for civilization regardless of politics and geography. More and more engineers and professionals are planning and building careers revolving around harnessing and domesticating energy resources as efficiently and effectively as possible. With the rapid development of power-hungry technologies that caused energy footprints of individuals across the globe to also increase at unprecedented rates, the need to supply and store energy to follow this demand has become paramount. Just think of the number of portable and mobile devices one has to charge and recharge, sometimes more than once per day: tablets, laptops, cell phones, etc. The convenience these devices offer in everyday life made them an integral part of it, to order food or a ride, to book tickets, to control remote processes, or to run businesses, the list of human activities that are centered around power-hungry devices just goes on and on.

Energy storage is an important facet of such development. And although storage technologies has existed as soon as energy supply had discrepancies with consumer demand, the available methods for energy storage have not enjoyed the same momentum in research and development as energy generation/transmission technologies. Chemical batteries and perhaps pumped hydro storage were the earliest and most trusted methods for energy storage, but unfortunately in today's world, they may not be flexible enough. The level of complexity in terms of the required portability and magnitude to respond to recent applications, be it domestic or industrial has necessitated a change in paradigm regarding what technologies can be used, and what are clearly outdated or obsolete. The suitability of the energy storage technology for the energy source, that would require the minimum conversion steps, has become a deciding factor in choosing storage for a project of any scale, as it directly determines the attainable efficiency with the least inevitable losses.

Being a founding member of the Sustainable and Renewable Energy Engineering Department at the University of Sharjah in 2012, I have faced unusual challenges in furnishing one particular lab: The Energy Storage Laboratory. The department teaches classical courses adapted from the mechanical engineering and electrical engineering disciplines, with the intuitive focus on courses necessary for a proper undergraduate background in subjects that deals with energy conversion and utilization. For example, the mechanical side of the specialty teaches thermodynamics, heat transfer, fluid mechanics, solar thermal systems, and wind energy. The electrical side introduces electrical power and machines, power electronics, and solar PV systems, among others. The lab sections were equipped with classical kits one would buy off-the-shelf, and the task was a relatively easy one. But for the "Storage Lab" things were quite different. There was no single supplier who was capable of suggesting a source to provide the equipment that would be as unique as the lab itself. Apart from introductory kits that were created to introduce energy concepts in general, energy storage (that was not a rechargeable chemical battery system) was not the specialty of anyone. We had to take matters into our own hands, and with the help of our amazing students, lab engineers, and local workshops, we were able to produce rudimentary devices for latent heat storage testing, modular compressed-air storage system, a buoyancy experiment (with buoys immersed in a plexiglass tank that not only endured the staggering water weight, but also the consistent jokes about adding some fish to it), and a Pelton turbine to emulate a pumped hydro storage system. Throughout the years, this lab was used to bring home important concepts of energy storage, with an important emphasis on debunking the exclusiveness of batteries as the sole energy storage solution. A pat on the back was delivered by an international accreditation team visiting the lab as part of the program reaccreditation, where they commended the effort put into making the equipment and delivering the scientific content. The lab has received several equipment upgrades this year, and we are thrilled to be home to such a pilot facility that has enriched the experience of our graduates and helped generate significant scientific research work.

This book was inspired by this local success story. It provides a stepping stone for students and professionals alike wishing to have the basics of mechanical energy storage technologies at their fingertips. The book bridges the gap that was widened by the focus on chemical storage and batteries. Although the techniques described in the chapters to come can be coupled with any energy resource, regardless of its rating and magnitude, there is an emphasis on the coupling with renewable resources. A fundamental mathematical background is given in each chapter, to attempt to keep the reader from continuously consulting other texts seeking fundamental basics. The book is written with fourth-year/graduate engineering students in mind, drawing from a decade-long experience in the subject matter. This being said, I am sure that energy and engineering professionals alike would find the information in the book quite useful as a reference when they are planning their next energy storage feat.

Sharjah, United Arab Emirates   Abdul Hai Alami, Ph.D.

Contents

About the Author

Dr. Abdul Hai Alami is an Associate Professor at the Sustainable and Renewable Energy Engineering Program at University of Sharjah.

He has received his Ph.D. from Queen's University in Kingston, Canada in 2006. Since then, he had held the position of Assistant Professor of Mechanical Engineering at the Hashemite University in Jordan, till he moved to the UAE in 2010 where he worked as a Mechanical Engineering Faculty at the Higher Colleges of Technology, Al Ain.

The current area of interest of him is the synthesis and characterization of mesoporous materials for third-generation photovoltaic solar cells, solar thermal energy utilization and augmentation (selective solar absorbers, evaporative cooling of PV modules), and novel ways of mechanical energy storage (buoyancy force, superconductor synthesis and deposition as well as compressed air).

List of Figures

List of Tables

List of Equations

Introduction to Mechanical Energy Storage

1.1 Introduction to Mechanical Energy Storage

This book will focus on energy storage technologies that are mechanical in nature and are also suitable for coupling with renewable energy resources. The importance of the field of energy storage is increasing with time, as the supply and demand cycles become more and more stochastic and less predictable. To complicate matter further, not only does the storage problem involve purely technical aspects, it surpasses it into an intermingled financial, political, and socioeconomic factors. It is also quite inevitable that storage professionals will be in high demand, whose background would include various intertwining areas of science and engineering, with the purpose to design and implement customizable storage options that responds to individual parameters and inputs. A classical mechanical, chemical, or electrical engineer may not fit the bill anymore, given the interdependence of energy conversion steps on the availability of new materials and new technologies that can be utilized for energy storage purposes.

This being said, the discussions and cases studies of this book will be directed at mechanical energy storage technologies. Naturally, this does not omit the importance of other types of technologies dealing with energy transmission and conversion but rather places a greater emphasis on technologies that can alternate between a charged state (high potential energy) and discharged state (high kinetic energy). This is a subfield that has developed and grown in past decades, yet has not received the deserved attention in the storage literature, nor have some novel technologies (e.g., work of a buoyancy force) been discussed in detail in other textbooks.

1.2 Need for Storage Technology

Power plants have always been designed to supply a certain average demand called the baseload. And while the supplied load level can be controlled to be reliably constant throughout the day, changes and fluctuations in demand from the consumer side impose many challenges on power plant operators. The answer to the question of how to supply inevitable demand peaks that occur when activities of the population require the extra draw on the power resources is not a straightforward one. Figure 1.1 shows a typical demand cycle that starts at a low level in early morning hours, surges upward around midday and early evening hours, and then declines toward late night, only to repeat itself the next day.

The area under the curve marked in red is the extra demand that the power utilities have to supply. If not, these utilities could find themselves operating near their maximum capacity mark, which may cause a blackout if at some point the plant rated power is exceeded. Worse, residents could live through a repeat of the massive blackout that occurred in northeastern United States (and southwestern Canada) in the hot summer of 2013, when higher than usual demand on one power plant caused it to go offline, causing the electrically tethered power plants to automatically jump into the rescue and direct their supply to the fallen part of the network and cause a domino effect of power failures that left the most important economic and industrial hub in the world without power and water for more than 72 h.

Unfortunately, the response to such problems is not a simple matter of building more power plants. Apart from the environmental issues and human health risks that arise whenever a new utilities project is erected, these major

© Springer Nature Switzerland AG 2020
A. H. Alami, *Mechanical Energy Storage for Renewable and Sustainable Energy Resources*,
Advances in Science, Technology & Innovation, https://doi.org/10.1007/978-3-030-33788-9_1

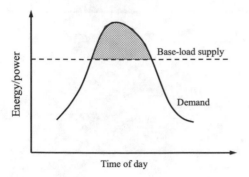

Fig. 1.1 Generic demand and supply of power over the span of a day

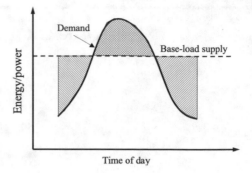

Fig. 1.2 Green areas where storage is permissible to keep power plant operating at baseload

projects require colossal budgeting from governments whom may not have the large sums required to invest in such a long-term solution. Also, building power plants that are capable of continuously supplying peak loads is seen as a waste of money and resources, because they respond to a fleeting surge in demand, only to have that extra capacity either idle or shut down more than half of its operational life. For this reason, expensive, yet temporary solutions in the form of gas turbines or internal combustion engines that can come online almost instantaneously are usually called upon to supply the demand transients, and then be shut off almost as fast as they were fired up.

And with the advent of reliably operational renewables, along with governments' mandates to reduce carbon dioxide emissions caused by the burning of fossil fuels, peak loads have found their match with technologies that are intensively available when those peaks occur. For example, solar photovoltaics and concentrated solar thermal power plants (CSP) are now augmented into the supply grid and their production mainly targets that red peak shown in the figure. For the moment they are not reliable enough to supply the baseload on their own, as their supply availability (the sun) is not always correctly predicted or ensured, but they definitely provide the extra energy needed to supply peak demand.

The missing link in the supply–demand cycle is storage. The proper selection, installation, and operation of storage technologies match the power/energy source in the required application. This means that whenever the supply is abundant, storage systems are charged and ready. Whenever demand supersedes supply, the power plant supplies its baseload, and the stored energy/power is withdrawn from its storage and is discharged to assist in resolving the strain on the power plant caused by peak load.

The green areas in Fig. 1.2 indicate where storage technologies need to be active. These areas will store the utilities supply, while their power plants operate at baseload all the time. This is more economical and efficient for conventional power plants, be it nuclear or fossil fuel based.

The book will also assist in answering the question about the selection of the most appropriate storage technology for a

given source of energy. It is well known from thermodynamics that more energy conversion steps will result in a decrease in the available energy, as each energy converter has a certain efficiency that it operates at. Image that one has three energy conversion devices, operating at an unattainably high efficiency of 90% (most energy conversion devices operate at far less efficiency). The overall efficiency of these devices in series would be $0.9 \times 0.9 \times 0.9 = 72.9\%$. This is such a waste, and thus it is better to decouple the electric generators from their driving steam turbines and reduce steam generation in the power plant during times of low demand than to continue generating electricity only to store it in batteries for later use.

1.3 Storage for Renewable Energy Resources

Devising storage technologies for renewable energy resources in particular is a crucial step in the design and implementation of any renewable energy facility. The storage has twofold importance in such installations since they need to respond to the uncertainty of both the demand and the supply. For example, the unpredictability of the availability of wind energy necessitates the full-scale operation of wind turbines whenever wind energy is available regardless of the demand level. However, without proper storage provisions, this energy could be wasted. The "proper" storage provision in this case is a technology that requires least energy conversion steps, which definitely rules out chemical batteries: imagine, with the help of Fig. 1.3, the losses incurred when converting the incoming kinetic energy of a wind stream into rotational energy in the turbine blades, then mechanical rotation of a generator that has friction and eddy losses, and then the thermal and chemical losses in the batteries.

Not omitting the losses that would also have to occur when converting the DC power of the battery into AC, stepping it up in transformers in order to transmit it to the grid also would cause transmission losses.

Fig. 1.3 Expected major losses for a wind turbine with chemical battery storage

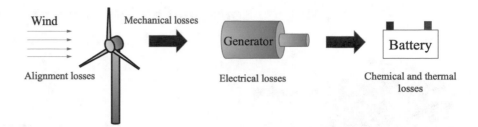

1.4 World Energy Resources and Consumption

Although a major worldwide drive has been initiated to limit the reliance on the burning of fossil fuels, the high calorific value of such fuels (especially natural gas), the mature technology of drilling, shipping, and distribution as well as the full dependence of all modes of transport on such fuels have made any initiative to change the consumption habits a painfully slow process.

In the United States, although a diverse set of resources are available to contribute to the energy mix, oil and gas have around 67% of the share of the energy consumed in 2018. The statistics are very similar in other countries around the world (Fig. 1.4).

It is also interesting to note that 38% of these resources are directed toward electricity generation, while the rest is used directly by industrial or residential users or for transportation applications. Even more interesting is that 25% of the energy is lost in the form of electrical system energy losses, leaving only 13% of useful electricity generated, as shown in Fig. 1.5.

The whole world is thus "plugged", which means that individuals prefer electrical energy over any type of energy form. The electrical consumption footprint of each individual on earth is increasing every year (just think of how many devices one is charging every day, perhaps more than once a day), and the trend does not seem to be easing off. The need for energy, especially in its refined form that pours out of wall outlets and plugs, is reaching an all-time high all around the world and with it, the need for energy portability,

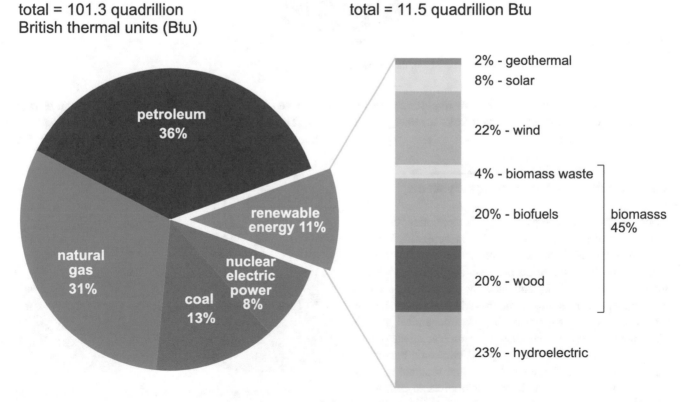

Fig. 1.4 U.S. energy consumption in 2018 classified by energy resource [1]

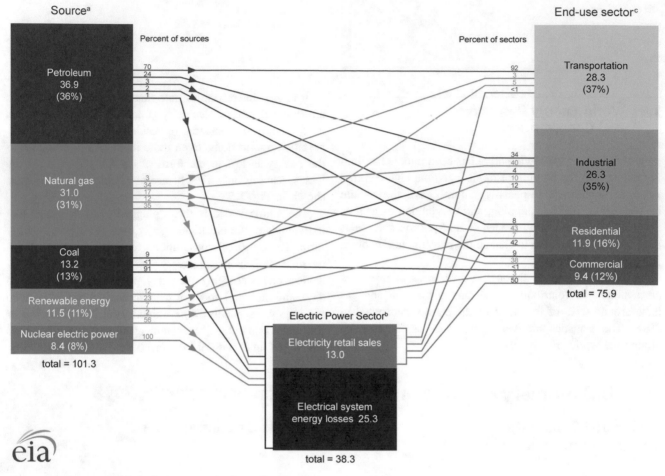

Fig. 1.5 Energy consumption by sector and resource in the United States in 2018 [1]

efficient energy transmission and more importantly energy storage is one of the most important tasks for energy engineers and professionals.

1.5 Classification of Storage Systems

Whenever the term "energy storage" is mentioned, the image of a chemical battery is almost always conjured. Batteries have been the de facto storage resort for decades, perhaps propelled by the reliance on the lead–acid batteries in early automotive applications, and also by the ease of stacking them up and connecting them in series or in parallel (as battery banks) to provide reliable electrochemical storage. The chemical storage concept is very simple: a battery consists of two electrodes made of good electronic conductors and immersed in an ionic electrolyte that separate them, as shown in Fig. 1.6. This is known as a galvanic cell, where the chemical force of ionic motion within the electrolyte must

be balanced by the flow of electrons in the external electric wires. This mechanism, of course, does not apply to the primary, non-rechargeable batteries category, like the AA

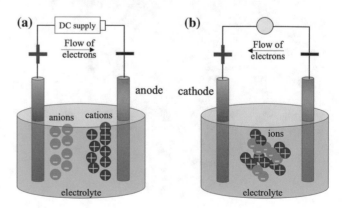

Fig. 1.6 Schematic of battery operation during **a** charging and **b** discharge

and AAA alkaline batteries we use for electrical appliances, as this variety cannot be recharged once they are depleted.

Figure 1.6 depicts the cycle of charging and discharging the battery. Once an external DC power is applied (Fig. 1.6a), the ions arrange themselves within the electrolyte next to their oppositely charged electrode (cations congregate close to the anode, while anions do the same around the cathode). Once discharge is allowed (a load is connected and shorts the electrical circuit as shown in Fig. 1.6b, the motion of ions in the electrolyte now is the opposite of that of the electrons in the wires and this continues until the chemical potential depletes by ending up with randomly dispersed ions in the electrolyte.

For the galvanic cell shown, the energy balance is between the ionic motion in the electrolyte (chemical potential) and the electrostatic electricity in the wire. A special form of Gibbs free energy (more about that in the next chapter) is related to the electrical potential as follows:

$$\Delta G^{\circ}_{\text{cell}} = -nFE_{\text{cell}} \tag{1.1}$$

where $\Delta G^{\circ}_{\text{cell}}$ is the tabulated value of Gibbs free energy of the electrolyte (taking into account both solvent and solute materials), n is the number of species in the electrolyte, F is Faraday's constant (96,485 C mol^{-1}), and E is the cell potential in volts. This equation can easily be rearranged to solve for the cell potential as $E_{\text{cell}} = -\frac{\Delta G^{\circ}_{\text{cell}}}{nF}$.

The basis for the galvanic cell is a redox reaction that is broken down into two half-reactions: oxidation taking place at the anode, where there is a loss of electrons, and reduction occurs at the cathode, where there is a gain of electrons. Electricity is generated due to the electric potential difference between the electrodes. The free energy term, $\Delta G^{\circ}_{\text{cell}}$, is a measure to how easily the oxidized species gives up electrons and how badly the reduced species wants to gain them and thus is dependent electrolyte selection.

For example, lead–acid batteries consist of electrodes made up of elemental lead (Pb) and lead oxide (PbO$_2$) immersed in an electrolyte of dilute sulfuric acid (H$_2$SO$_4$). During discharging, few changes occur in galvanic cell, such as the electrolyte conversion into water and the electrodes into lead sulfate (PbSO$_4$) due to the interaction with the sulfate ions. The efficiency of lead–acid batteries ranges between 70 and 90%, and their cost is relatively low (US $200–400 per kWh). These batteries are mostly used in uninterruptible power supply (UPS) systems of the capacity of 100 s Ampere hour (Ah) range. The other advantage of these batteries is they keep the cell voltage steady (~ 2 V) till around 20% of the original fully charged capacity, unlike the more recent lithium–ion batteries where the cell voltage falls steadily as the battery is discharged, as shown in Fig. 1.7.

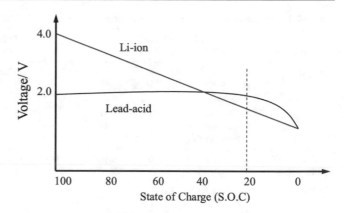

Fig. 1.7 Battery cell voltage versus state of charge for lead–acid and Li–ion batteries

The issue, however, remains with the lack of cycle efficiency and energy density as their life cycle between 500 and 1000 cycles and energy density of 50–90 Wh/L. The energy density is high due to the nature of the lead electrodes supporting high energy density operation. The battery performance becomes poor with temperature changes (which was a major issue in starting your car after summer vacation); therefore, a dedicated cooling system is required in large battery bank installations for them to operate at optimal efficiency.

An interesting listing that shows various battery types, electrodes, electrolytes, and operational voltage is shown in Table 1.1 [2].

1.6 Available Storage Technologies

Luckily, batteries are not the only option for energy storage. It is true that electricity is the most sought-after form of energy, but this does not mean that only electrochemical storage routes are the most suitable ones. A general classification is shown in Fig. 1.8.

Each of the technologies presented in Fig. 1.8 can be matched with an energy resource and utilized at a certain stage of the energy conversion cycle. It is quite common to implement more than one storage technique whenever the need arises and the technical feasibility in terms of overall system efficiency permits.

The worldwide energy storage reliance on various energy storage technologies is shown in Fig. 1.9, where nearly half of the storage techniques are seen to be based on thermal systems (both sensible and latent, around 45%), and around third of the energy is stored in electrochemical devices (batteries).

The progress with chemical storage in batteries has been rapid, especially with the introduction of lithium–ion (Li–ion) batteries to replace the more outdated and

Table 1.1 Characteristics of most common battery systems and their commercial application

Battery	Reactions and electrolyte	Cell voltage (E_0), V	Commercial unit capacity and location
Lead–acid	Anode: $Pb + SO_4^{-2} \rightarrow PbSO_4 + 2e^-$ Cathode: $PbO_2 + 4H^{+1} \rightarrow PbSO_4 + 2H_2O$ Electrolyte Diluted H_2SO_4	2.1	10 MW/40 MWh in 1988 California-Chino load leveling 300 kW/580 kWh Turn key system load leveling 14 MWh in 1986 BEWAG Plant Berlin, Germany 4 MWh/1 h Madrid, Spain 14 MWh/1.5 h PREPA, Puerto Rico
Nickel Cadmium	$2NiO(OH) + Cd + 2H_2O \rightarrow Ni(OH)_2 + Cd(OH)_2$	1.2	27 MW/6.75 MWh GVEA Alaska Var compensator
Sodium sulfur	Discharging reaction: $2Na + 4S \rightarrow Na_2S_4$ Electrolyte: beta aluminum	2	9.6 MW/64 MWh Japan (Voltage sag peak load shaving) 6 MW/8 h Tokyo electric power company 8 MW/7.25 h Hitachi Plant Ohio US 1.2 MW demonstration
Lithium–ion	Anode: $LiC_6 \rightarrow Li^{+1} + e^{-1} + C_6$ Cathode: $CoO_2 + Li^+ + e^{-11} \rightarrow LiCoO_2$ Electrolyte: Lithium salts in organic solvent	3.6/3.85	Mobile and computer application
Nickel–metal hydride	Anode: $Ni(OH)_2 + OH^{-11} \rightleftarrows Ni(OH) + H_2O + e^{-11}$ Cathode: $H_2O + M + e^{-11} \rightleftarrows OH^{-11} + MH$ Electrolyte: Potassium hydroxide		
Metal–air battery	Anode: $Zn + 4OH^{-11} \rightarrow Zn(OH)_4^{-2} + 2e^{-11}$ In Fluid: $Zn(OH)_4^{-2} \rightarrow ZnO + H_2O + 4OH^{-11}$ Cathode: $O_2 + 2H_2O + 4e^{-11} \rightarrow 4OH^{-11}$ Overall reaction: $2Zn + O_2 \rightarrow 2ZnO$ Electrolyte: KOH/KOH with solid polymer membrane	1.65 for Zn Air	Metal as fuel and Air as oxidant 1 MWh Con Edison, National Grid, USA

environmentally hazardous lead–acid ones. The energy density of the Li–ion battery is three times that of their lead–acid counterparts, mainly due to the low atomic mass of 6.9 μ of Li as compared to 207 μ for lead, which is also better for the ionic diffusion process. The Li–ion cell voltage is also higher than that of lead–acid batteries (3.6 V for the former compared to 2.1 V for the latter). This means that the cell size and cost is reduced once the cells are connected in series or parallel. The self-discharge rate of Li–ion batteries is around 5% per month, which means that an idle battery would be 60% depleted within a year. The battery life is up to 1500 cycles, and its lifetime is temperature dependent, which means that the aging phenomenon (the inability to reach 100% of the original battery capacity) is much faster at higher temperatures and the lifetime of the battery is reduced due to deep discharge. These factors limit the use of Li–ion batteries for such applications where the complete discharge is frequent. But the Li–ion batteries are still considered to be the future of chemical portable storage, be it for mobile and personal computing, as a main storage option of photovoltaic installations, or for automotive application like the example of Li–ion batteries in the Tesla Roadster. Tesla is currently the largest manufacturer of fully electric vehicles (EV).

The Tesla Roadster is powered by lithium–ion (Li–ion) battery system, which is comprised of 6,831 individual Li–ion cells, taking up the storage in the trunk and weighs about 900 lb. The battery system is the secret behind the 4-s 0–60 mph acceleration and high driving range. Tesla Li–ion battery system in the Tesla Roadster represents has the most advanced, commercially available battery technology. The Li–ion batteries on the vehicle perform better than nickel–metal hydride (NiMH) cells and lead–acid cells found in other electric vehicles (Fig. 1.10)

Fig. 1.8 General classification of available storage technologies

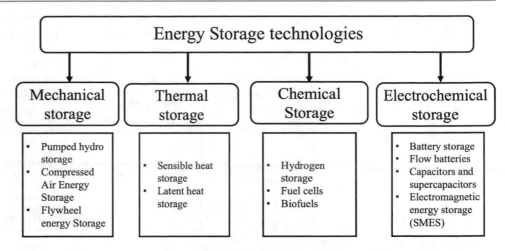

Energy Storage technologies

Mechanical storage
- Pumped hydro storage
- Compressed Air Energy Storage
- Flywheel energy Storage

Thermal storage
- Sensible heat storage
- Latent heat storage

Chemical Storage
- Hydrogen storage
- Fuel cells
- Biofuels

Electrochemical storage
- Battery storage
- Flow batteries
- Capacitors and supercapacitors
- Electromagnetic energy storage (SMES)

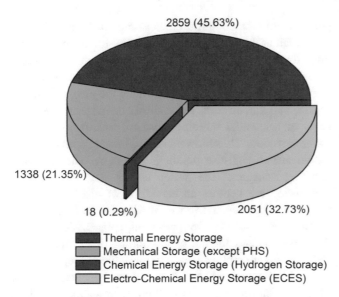

2859 (45.63%)

1338 (21.35%)

18 (0.29%)

2051 (32.73%)

■ Thermal Energy Storage
■ Mechanical Storage (except PHS)
■ Chemical Energy Storage (Hydrogen Storage)
■ Electro-Chemical Energy Storage (ECES)

Fig. 1.9 Grid-connected operational capacities of all storage technologies [2]

Fig. 1.10 A schematic showing the Tesla Roadster and the battery carried in back of the vehicle [3]

Tesla reports that one of the most difficult challenges in battery design is increasing energy density while also maximizing battery life span. Li–ion chemistries have achieved better combinations of these parameters than other battery technologies. Yet, there is still a trade-off between energy and life, even within the family of Li–ion.

1.7   Mechanical Storage Systems

The storage branch that is the focus of this book is mechanical technologies of energy storage. In Fig. 1.8, the classification shows that mechanical systems are strictly those who have a distinct and clear conversion of potential and kinetic energies. By examining the available reports in literature, it is clear that other technologies can be added to the original classification, as shown in Fig. 1.11.

The definition of mechanical storage technologies can also be expanded to include thermal storage systems, as it can be argued that the thermal storage mechanism in any material is based on a molecular-level increase in kinetic (vibrational) energy, which eventually leads to microstructural changes once the latent heat necessary to alter the phase of a material is reached. Although not present in Fig. 1.11, this book has a chapter devoted to thermal energy storage (TES) systems.

The common traits of mechanical storage technologies are their technical simplicity, robustness, and economic feasibility. And just like electrical energy being the current ultimate goal for all energy generation and conversion activities, mechanical energy is required to rotate the electrical generators that are connected to the grid, as shown in Fig. 1.12. This is a strong motivation to implement mechanical storage systems as early as possible in the power generation cycle.

Fig. 1.11 Classification of mechanical energy storage technologies

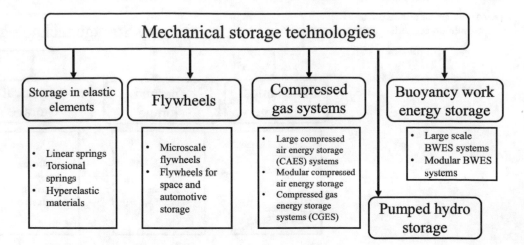

Note that since energy storage in batteries can only happen after the electrical generator and before the grid connection, excessive conversion losses will be incurred due to attempting to wait till electricity is produced to store energy. It is thus beneficial to design and operate storage technologies that can absorb the kinetic energy and then release it to the generators whenever needed. Most of the technologies presented in Fig. 1.11 are at a level of maturity and commercial availability that they present a viable alternative to battery storage.

Mechanical technologies such as pumped hydro storage provide an enormous potential for load leveling as they can absorb all load abnormalities (be it supplying or storing the energy load). Their concept of operation has been applied on a small scale in water distribution in the past decades in countries with no skyscrapers. Figure 1.13a shows the simple principle of operation of such tank, where water is pumped from the central utilities storage tanks into the elevated tank at the top of the tower during low demand times (nighttime or weekends). Once water distribution is required (mainly at rush hour), the water is allowed to flow downward by gravitational pull, transforming the potential energy ($\rho g \Delta h$) into kinetic energy that allows the water to reach the tanks of residential buildings that are usually placed at a lower elevation. These towers are no longer in service with the advent of high-rise buildings and more efficient water pumping schemes. The water towers that did not get demolished are kept as an exhibition of discontinued old technology (see Fig. 1.13b for Al Khazzan Park, Dubai).

This book will start by presenting some basic concepts that are important for the understanding of the subsequent chapters. An attempt was made to make the concepts in each chapter self-contained, but it is always beneficial for the reader to consult textbooks of fluid, thermodynamics, heat

Fig. 1.12 Reliance of electrical power generation on rotational motion

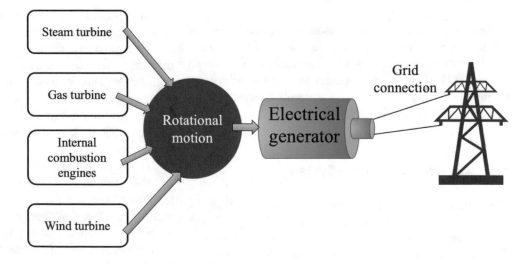

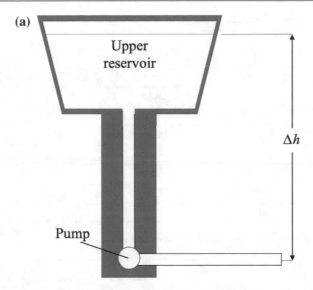

Fig. 1.13 **a** Water tower principle of operation and **b** an old water tower in Dubai, United Arab Emirates that received the status of a historical monument (Al Khazzan Park, Dubai)

transfer, and materials whenever a deeper understanding of a concept is needed. The third chapter introduces energy storage in elastic (and hyperelastic) materials and springs. Chapter four is devoted to thermal energy storage in its sensible and latent forms, along with some application examples in concentrated solar plants (CSP). Chapter five discusses flywheel technologies and implementation in automotive and space applications. Chapter six contains discussions of pumped hydro storage technologies and its

place as the "largest battery system" in terms of energy capacity. Chapter seven is for compressed-air energy storage (CAES) both large scale and small scale, and also compressed gas energy storage (CGES) systems are also introduced and the potential of their utilization is presented and discussed. Chapter eight introduces the concept of buoyancy work energy storage (BWES) and its application for remote and maritime applications that shows a great potential. The last chapter presents a brief discussion on recent

Table 1.2 List of prefixes, SI units, and conversion factors [4]

(a) Prefixes for powers of 10

Fractional order			Multiple order		
Prefix	Symbol	Value	Prefix	Symbol	Value
Atto	a	10^{-18}	Kilo	k	10^3
Femto	f	10^{-15}	Mega	M	10^6
Pico	p	10^{-12}	Giga	G	10^9
Nano	n	10^{-9}	Tera	T	10^{12}
Micro	μ	10^{-6}	Peta	P	10^{15}
Milli	m	10^{-3}	Exa	E	10^{18}

(b) Basic SI units

Unit	Name	Symbol
Length	Meter	M
Mass	Kilogram	Kg
Time	Second	S
Electric current	Ampere	A
Temperature	Kilogram	K
Luminous intensity	Candela	cd
Plane angle	Radian	Rad
Solid angle	Steradian	sr
Amount	Mole	mol

(c) Derived SI units

Unit	Name	Symbol	Definition
Energy	Joule	J	$kg\,m^2 s^{-2}$
Power	Watt	W	$J\,s^{-1}$
Force	Newton	N	$J\,m^{-1}$
Electric charge	Coulomb	C	$A\,s$
Potential difference	Volt	V	$J\,A^{-1}s^{-1}$
Pressure	Pascal	Pa	$N\,m^{-2}$
Electric resistance	Ohm	Ω	$V\,A^{-1}$
Electric capacitance	Farad	F	$A\,s\,V^{-1}$
Magnetic flux	Weber	Wb	$V\,s$
Inductance	Henry	H	$V\,s\,A^{-1}$
Magnetic flux density	Tesla	T	$V\,s\,m^{-2}$
Luminous flux	Lumen	lm	$cd\,sr$
Illumination	Lux	lx	$cd\,sr\,m^{-2}$
Frequency	Hertz	Hz	$cycle\,s^{-1}$

(d) Conversion factors

Type	Name	Symbol	Approximate value
Energy	Electron volt	eV	$1.6021 \times 10^{-19}\,J$
Energy	Erg	erg	$10^{-7}\,J$ (exact)
Energy	Calorie (thermochemical)	cal	$4.184\,J$
Energy	British thermal unit	Btu	$1055.06\,J$

(continued)

Table 1.2   (continued)

(d) Conversion factors

Type	Name	Symbol	Approximate value
Energy	Q	Q	10^{18} Btu (exact)
Energy	Quad	q	10^{15} Btu (exact)
Energy	Tons oil equivalent	toe	4.19×10^{10} J
Energy	Barrels oil equivalent	bbl	5.74×10^{9} J
Energy	Tons coal equivalent	tce	2.93×10^{10} J
Energy	Cubic meter of natural gas	–	3.4×10^{7} J
Energy	Kilogram of methane	–	6.13×10^{7} J
Energy	Cubic meter of biogas	–	2.3×10^{7} J
Energy	Liter of gasoline	–	3.29×10^{7} J
Energy	Kilogram of gasoline	–	4.38×10^{7} J
Energy	Liter of diesel oil	–	3.59×10^{7} J
Energy	Kilogram of diesel oil/gasoil	–	4.27×10^{7} J
Energy	Cubic meter of hydrogen at 1 atmospheric pressure	–	1.0×10^{7} J
Energy	Kilogram of hydrogen	–	1.2×10^{8} J
Energy	Kilowatt-hour	kWh	3.6×10^{6} J
Power	Horsepower	hp	745.7 W
Power	Kilowatt-hour per year	kWh/y	0.114 W
Radioactivity	Curie	Ci	$3.7 \times 10^{8} s^{-1}$
Radioactivity	Becquerel	Bq	$1 s^{-1}$
Radiation dose	Rad	rad	10^{-2} J kg^{-1}
Radiation dose	Gray	Gy	J kg^{-1}
Dose equivalent	Rem	rem	10^{-2} J kg^{-1}
Dose equivalent	Sievert	Sv	J kg^{-1}
Temperature	Degree Celsius	°C	K − 273.15
Temperature	Degree Fahrenheit	°F	9/5 °C + 32
Time	Minute	min	60 s (exact)
Time	Hour	h	3600 s (exact)
Time	Year	yr	8760 h
Pressure	Atmosphere	atm	1.013×10^{5} Pa
Pressure	Bar	bar	10^{5} Pa
Pressure	Pounds per square inch	psi	6890 Pa
Mass	Ton (metric)	t	10^{3} kg
Mass	Pound	lb	0.4536 kg
Mass	Ounce	oz	0.02835 kg
Length	Ångström	Å	10^{-10} m
Length	Inch	in	0.0254 m
Length	Foot	ft	0.3048 m
Length	Mile (statute)	mil	1609 m
Volume	Liter	l	10^{-3} m^{3}
Volume	Gallon (US)	gal	3.785×10^{-3} m^{3}

technologies and applications of the various mechanical storage technologies discussed in this book.

1.7.1   Units and Conversions

With the many standards devised for energy and energy conversion throughout the years, it would be beneficial to present a list of such standards, prefixes, and conversion factors in a single location. This information is available in Table 1.2.

References

1. U. S. E. Information Administration, U.S. energy facts explained (2018). https://www.eia.gov/energyexplained/us-energy-facts/
2. N. Khan, S. Dilshad, R. Khalid, A.R. Kalair, N. Abas, Review of energy storage and transportation of energy. Energy Storage **1**(3) (2019)
3. M. Eberhard, A bit about batteries. https://www.tesla.com/fi_FI/blog/bit-about-batteries
4. B. Sørensen, *Renewable Energy Conversion, Transmission, and Storage*. Elsevier/Academic Press (2007)

2.1 Introduction

Whenever energy conversion is sought, especially involving thermal systems, thermodynamics and its laws are called upon for help. These laws also aid us in quantifying the amount of losses that are inescapable during the operation of a system under nonideal conditions, so we can design more realistic processes.

This chapter will introduce the concepts of conservation of energy and efficiency of some thermal and mechanical systems based on basic principles of thermodynamics, heat transfer, gas dynamics, and rigid body motion. The energy transfer required for phase changes within a compound will also be briefly discussed by introducing phase diagrams.

2.2 Basic Energy Conversion Concepts

The final useful energy form is usually different from the original form the energy initially assumed. For example, electrical energy is the most common final energy form that has been standardized across the globe, no matter whether it is produced from the oxidation of chemical energy in fossil fuels (combustion) or through photovoltaic conversion. To calculate the efficiency of any energy conversion device, it is necessary to use the same yardstick to measure and compare inputs and outputs. Also, it is extremely useful to know that, for example, the area under the stress–strain diagram of a material is the energy per unit volume of the test specimen, or that the motion of a piston against a fluid under its own weight, which is a form of work, also has the units of energy.

The SI unit of energy is Joule, or (J), which is a derived unit from kg m^2/s^2. Power is the rate of doing work (or spending energy), or (J/t), is better known as Watt, or (W). Power has been classically related to horsepower, and 745.7 W equal 1 hp (see Table 1.2). This unit is popular when selecting and describing motor sizes for a particular application, or sources of shaft work.

2.2.1 Thermodynamics

The energy added to or extracted from a system is always associated with changes in macroscopic variables such as temperature, pressure, and volume. The system here refers to the specific substance we choose to perform the energy balance upon and is assumed to be enclosed within a fictitious enclosure we call the control volume. The four laws of thermodynamics are in place to predict the effect of the interaction of energy (in the form of either heat or shaft work) with a specific amount of a substance that makes up the "system" we are interested in. The system can either be closed or open. The former has a fixed amount of substance existing within a container with one of the following attributes: the size of the container is constant and remains so throughout the applied process (constant volume system) or the pressure applied on the system does not change while its volume does, or what is known as a "constant pressure" system. Both systems are depicted in Fig. 2.1.

The systems are "closed" such that none of the enclosed amounts of the substance under consideration is allowed to enter or leave through the imaginary control surface lines. Only energy in the form of heat or shaft work can cross the boundary and cause changes in temperature and either pressure or volume.

Open systems impose fewer restrictions on the flow of the substance in or out of its container, as long as we know the amount that enters or leaves per unit time. This task becomes especially easy if the flow does not change with time, also known as steady flow (Fig. 2.2).

The energy balance for each of these systems is made easy with the help of the second law of thermodynamics, also known as the conservation of energy principle. It predicts how the system will react to the energy input (either heat or work) by changing its pressure and temperatures.

2.2.1.1 Internal Energy

The concept of specific internal energy, u (J/kg K), refers to the sum of interatomic bond energies between the molecules

© Springer Nature Switzerland AG 2020
A. H. Alami, *Mechanical Energy Storage for Renewable and Sustainable Energy Resources*,
Advances in Science, Technology & Innovation, https://doi.org/10.1007/978-3-030-33788-9_2

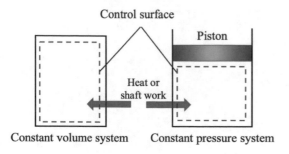

Fig. 2.1 Constant volume and constant pressure systems

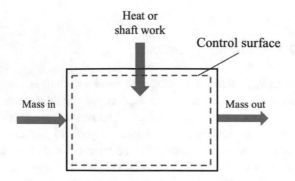

Fig. 2.2 Steady-flow, open system

that make up a certain material, as well as their kinetic and potential energies resulting from molecular vibrations of any known matter existing above 0 K. This can be summarized as

$$u = e + KE + PE \tag{2.1}$$

where e is the total bond energies, KE is the kinetic energy if the system is in motion, and PE is the potential energy if the system is located away from a certain reference frame.

The importance of this definition is that internal energy represents the system to which heat or shaft work is supplied. The direct relationship between the externally applied energy (heat or work) and the change in internal energy is described by the following form of the second law of thermodynamics:

$$\Delta u = q + w \tag{2.2}$$

where q is the applied heat and w is the shaft work.

By reviewing the constant volume enclosure of Fig. 2.1, we can see that the internal energy can be calculated based on the specific heat of the substance at constant volume, such that

$$\Delta u = c_v \Delta T \tag{2.3}$$

Equations (2.2) and (2.3) offer straightforward techniques of estimating the internal energy of a substance, but it should

be noted that texts of thermodynamics should be consulted for more accurate estimations of the internal energy.

2.2.1.2 Enthalpy

The above relationship works for a closed system with no changes in the dimensions of the control volume. However, any change in the control volume boundary indicates a different state of "compression" against the substance in the enclosure. Reexamining the control pressure schematic of Fig. 2.1, the weight of the constant cross-sectional area piston applies a pressure on the substance in the enclosure, changing the elevation of the piston (up or down) with respect to some original location. Since the work done by the weight of the piston along a certain distance, Δy, has the units of energy too, this gives rise to enthalpy, h, which is defined as

$$h = u + pv \tag{2.4}$$

where u is the internal energy, p is the pressure (piston weight per its area) in Pa, and v is the specific volume of the substance in the enclosure in m^3/kg.

And just like the internal energy, the enthalpy too can be estimated by evaluating the specific heat of the substance at constant pressure:

$$\Delta h = c_p \Delta T \tag{2.5}$$

The relationships for internal energy and enthalpy should be used in conjuncture with heat transfer relations that will be introduced in later sections of this chapter either to quantify the effect of heat transfer to or from the system, or to arrive at the required amount of heat needed to raise the system temperature a certain amount.

2.2.1.3 Entropy

The third law of thermodynamics puts restrictions on the directions energy can take to or from the system. Heat, for example, can only proceed from locations of high temperatures to low temperatures. The property designated for this task is called entropy, s, and it describes the amount of disorder a system possesses after an interaction with energy. Entropy is divided into two components: thermal and configurational. The former is the common thermodynamically defined concept of chaos that increases with added heat and is defined as

$$s_{th} = q/T \tag{2.6}$$

where s_{th} is the thermal entropy, q is the heat transfer, and T is the temperature (in K).

The configurational entropy, s_{conf}, on the other hand, relates to the microstructural order of the structure of the substance itself. Thus, a crystalline material is expected to have lower values of configurational entropy than an amorphous solid material, while the liquid phase of a material has higher values of s_{th} than its solid phase.

We can also define the total entropy, s, to be the sum of the thermal and configurational components:

$$s = s_{th} + s_{conf} \qquad (2.7)$$

2.2.1.4 Gibbs Rule and the Free Work Principle

For a storage system to be sized properly, it is imperative to be able to quantify the energy available within the system that can be harnessed at a later time. This is straightforward in the design of mechanical systems that are discussed in the next section, where the relationship between potential and kinetic energies can be determined, even in a nonideal system where losses are involved. However, whenever potential is disguised as a chemical or latent potential that pushes a chemical reaction or a phase change (thermal or metallurgical) in a certain direction, a more involved analysis is needed. This comes in the form of a relation presented by J. Willard Gibbs in 1873 that describes the driving force for any process in terms of the Gibbs free energy relation:

$$\Delta G = \Delta H - T\Delta S \qquad (2.8)$$

where H is the enthalpy, S is the entropy, and T is the process temperature. The extreme case of no free energy ($G = 0$) means that the system has no free energy to do any work, and thus $H = TS$, or the amount of total energy (H) is equal to the amount stored in the system, TS.

2.2.2 Dynamics of Rigid Bodies

The laws of dynamics govern the motion of particles or rigid bodies, be it a translation, rotation, or both. Dynamics has two main branches: kinematics and kinetics, where the former deals with position, velocity, and acceleration and the latter studies the forces that causes the position, velocity, and acceleration to change.

And just like laws of thermodynamics govern thermal systems, the three laws of motion developed by Newton are the terms by which the motion of a particle or a rigid body takes place. The simple term of Newton's second law ties up the kinetic and kinematic aspects of this motion, such as

$$\Sigma F = ma \qquad (2.9)$$

where ΣF is the resultant force applied and a is the resulting acceleration.

From this and for an ideal (no losses due to friction, windage, or others), the conservation of energy equation

between two points defined by a Newtonian reference can be written as follows:

$$KE = PE = \text{constant} \qquad (2.10)$$

where the kinetic energy, KE, and potential energy, PE, are measured from that reference and are defined as

$$KE = \frac{1}{2}mv^2 \qquad (2.11)$$

where v is the linear (translational) velocity of the particle (or center of mass of a rigid body). Even on a curvilinear path, this velocity will always remain tangential to that path.

$$PE = mgh \qquad (2.12)$$

where h is the height of a particle (or center of mass of a rigid body) above a certain datum and g is the acceleration of gravity.

Equations (2.11) and (2.12) are both scalar "point functions", which is very attractive in estimating the lossless mechanical energy exchange between two points of interest. This concept has been deployed in many applications, from amusement park rides to estimate the energy conversion required to the impact toughness tests of materials to determine the amount of energy a material can absorb to fracture.

The above cases are illustrated in Fig. 2.3, where a mass, m, starts motion from rest ($KE = 0$ and $PE = mgh_1$). The mass is supposed to reach an elevation from the datum line equal to its starting elevation ($h_1 = h_2$), unless a nonconservative force acts upon it at point 3 for a certain distance, generating a negative work (measured in units of energy, J) that will eat away its available energy and will cause it to reach the smaller height of h_2.

These concepts, although straightforward, are of immense practical use, since the energy deficit $\Delta h = h_1 - h_2$ sometimes should be compensated for by work done on the system in the form of a pumping "head" in the case of fluid flow in rough pipes, for example. This enables engineers to estimate the size of the motor that will drive the pump to achieve a final height of $h_1 = h_2$.

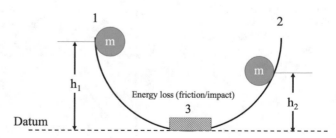

Fig. 2.3 Energy balance of a mass, m, rolling on a curvilinear path between points 1 and 2

2.2.3 Heat Transfer

In energy storage applications, heat as an energy form is conveniently divided into two forms: sensible and latent. Heat transfer occurs by one or more of three mechanisms: conduction, convection, or radiation.

2.2.3.1 Sensible Heat

Employing sensible in storage heat is accomplished by increasing the internal energy level of a certain mass of a material, thus creating a "sensible" temperature difference that can be estimated from Eq. (2.2). An example of this is boiling a certain amount of water (water has the highest value of internal energy at 4186 J/kg °C). If this water is stored in an insulated container, the energy in its sensible thermal form can be stored and used later, either directly or through heat transfer in a heat exchanger. Other examples of sensible heat utilizations in storage are the heating of thermal bricks using hot plates during times where the demand on electricity is low (in case there is an incentive plan of tariff reduction) and using them for space heating during times where the demand, and consequently the price, of electricity is high. This rudimentary storage technique is based on load shifting that affects the demand/supply continuum and causes a more homogenous utilization of the available energy, as discussed in Chap. 1. An example of such load shifting is shown in Fig. 2.4 for the simple case of sensible heat storage in thermal bricks. Thus, by preheating the bricks, the demand peak is partially transferred to a time the supply is capable of covering the demand, driving the demand/supply cycle into better stability and control.

2.2.3.2 Latent Heat

The concept of latent heat is synonymous to phase change in materials, as it is defined as the amount of energy per unit mass required to change the phase of the material. It occurs at constant temperature according to Fig. 2.5, and hence it has the intrinsic ability to store thermal energy at an energy level equivalent to the x-distance between the phases.

Depending on the size and application, the choice of a phase-change material for energy storage should usually

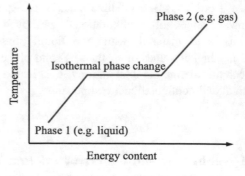

Fig. 2.5 Latent energy addition/rejection at constant temperature

couple a high value of specific heat of fusion with a low melting point temperature.

2.2.4 Phase Diagrams

Another resource of energy storage can happen through phase transformations, especially ones that are not intuitive as ones pertinent to latent heat. For example, α-iron has a body-centered cubic (BCC) unit cell at room temperatures, but above 914 °C the γ-iron has a face-centered cubic (FCC) unit cell right up to 1391 °C, at which δ-iron prevails, which is again BCC, and stays this way till iron eventually melts at around 1536 °C.

Although thermal energy storage potential is limited in magnitude and constrained by available technology, nanoscale applications and devices can benefit from enhanced heat removal and perhaps storage.

In order to determine the degrees of freedom, or the operational conditions and components properties (pressure, temperature, composition, etc.) required to describe the behavior of a thermal system under equilibrium conditions, Gibb's phase rule is used. The degrees of freedom, F, are determined by the number of phases, P, and components, C, present. The rule is written as $F = C - P + 2$, and the quantity F refers to the number of intensive thermodynamic or material parameter that must be specified in order to define the system and its properties.

Fig. 2.4 Load shifting by distributing the demand toward times of abundant supply and low demand

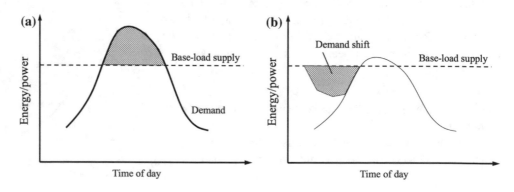

As an example, the binary (two-component, A and B) eutectic phase diagram shown in Fig. 2.6 has three single-phase regions and three two-phase regions. The eutectic point is characterized by the reversible reaction: *Liquid* $\rightleftharpoons \alpha + \beta$, at which the phase transformation goes directly from liquid to solid upon cooling and the reverse upon heating. This region, however, is not desirable for latent thermal heat storage as the inertia it offers is quite limited. From a storage perspective, the best composition lines should pass through the "mushy zone", where liquid exists in equilibrium with the solid solution (α) or (β) and the phase transformation (solid–liquid or liquid–solid) has to overcome the thermal inertia (the mass and specific heat of the material) over a longer period of time compared with compositions at the eutectic point. This is a good example of the role configurational entropy plays in energy storage, as the material changes phase with changes in both thermal and configurational entropies.

2.2.5 Compressible Flow

Most of the previous discussion in thermodynamics and heat does not take into account major changes in density caused by the ability of a material to be significantly compressed in response to small to moderate loads.

Under conditions of relatively low pressure and/or high temperature, the density is relatively low. The ideal gas law can be used to relate fluid properties according to the following equation:

$$p = \rho R T \tag{2.13}$$

where p is the absolute pressure (Pa), ρ is the density of the fluid (kg/m^3), T is the absolute temperature (K), and R is the

individual gas constant (N m/kg K). The latter is found by dividing 8314 by the molecular mass of the gas or can be found listed in tables.

In general, most processes involving gaseous substance can be represented by a polytropic process that follows the relation

$$p v^n = const = C_1 \tag{2.14}$$

where n is the polytropic exponent that can be any positive number. Some of the special cases for it are listed in Table 2.1.

The values in the table can also be conveniently visualized in Fig. 2.7 showing pertinent the p-v and T-s diagrams.

In compressible flow, the three main factors affecting the properties of the fluid are friction, heat transfer, and area changes. Generally, to analyze the flow problems through any storage device involving a compressible substance, only the most dominant effect of the three is considered. This helps the development of a simple solution within an acceptable range of accuracy, particularly in designing geometries that would decrease the likelihood of developing shockwaves that result from flowing at the speed of sound ($a = \sqrt{\gamma \cdot RT}$, where γ is the specific heat ratio, R is the gas constant, and T is the absolute temperature).

The adiabatic compressible flow is an example of neglecting heat transfer and shaft work, and thus eliminating any losses. This approach highlights area (geometry) changes as the main influence on energy conservation and conversion between the starting and end points of the fluid flow. In such cases, the energy balance usually takes place between two states. One is a state where the fluid velocity is negligible and maximum potential energy conditions prevail, called stagnation. The other state is characterized by fluid motion at a speed at or above the speed of sound and thus is measured as a fraction of it by introducing the Mach number. The formulation and solution of these relations will be introduced in Chap. 7 when compressed-air energy storage (CAES) systems are introduced and analyzed.

2.2.6 Electrical Generators

The importance of electrical energy as a terminal energy form, upon which converges all techniques of energy generation and conversion, is undeniable. And since mechanical energy storage techniques involve producing rotational motion in an air turbine, water turbine, flywheel, or a cable wound around a pulley (see again Fig. 1.12), electrical generators are an important component in the energy recovery step during the charge/discharge cycle, to directly transform kinetic energy into electrical.

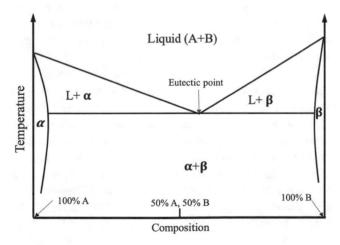

Fig. 2.6 Generic binary eutectic phase diagram

Table 2.1 Some values of *n* with their corresponding special case of the polytropic equation

$n = 0$	$P = $ constant
$n = 1$	$T = $ constant
$n = \gamma$	$s = $ constant
$n = \infty$	$v = $ constant

Fig. 2.7 Polytropic relation at different values of *n*

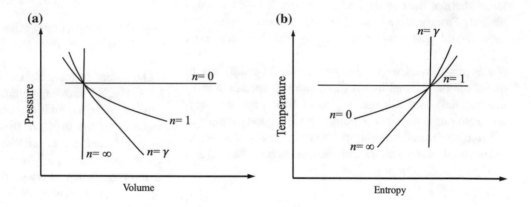

The trinity of magnetic field, electrical field, and rotational motion is the heart of electrical machines. Combine any two, and you get the third, and thus motors and generators are sometimes used interchangeably. As shown in Fig. 2.8, in the presence of electrical windings and a magnetic field, rotational motion will result in the generation of current within the windings (generator operation), while the passing of current within the windings in the presence of the magnetic field will cause rotational motion to happen (motor operation).

Depending on the type of electrical current sought, generators can be used to produce either direct or alternating current generators. Since the scope of this book does not allow a deep coverage of electrical machines, the main differences between the two types are listed in Table 2.2, with a depiction of each shown in Fig. 2.9.

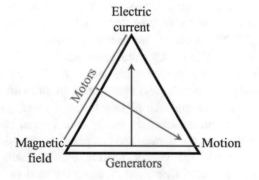

Fig. 2.8 Operational differences between generators and motors

2.2.7 Ragone Plots

The Ragone plots are important in classifying the energy charge/discharge capabilities of various storage systems. The energy density or power density concepts are as interrelated as the energy and power concepts themselves. Some applications require large amounts of energy to be expended over a short period of time, while others allow this energy to be utilized over a longer duration. The former is classified as power density applications, while the latter is energy density applications, with contemporary storage systems being able to cater for one of them in a better fashion than other applications. The usefulness of representing the properties of various storage systems in terms of a power–energy plane (or their densities) is evident. Since performance characteristics as well as the application requirements can be displayed in the same plot, it is more straightforward to understand any trade-off between high power and large available energy.

By considering Fig. 2.10 that depicts a Ragone plot for various energy storage technologies, it is not surprising to see capacitors occupying the high power, low energy density range of the plot, while lead–acid batteries residing at the other extreme of high energy, low power density. Note that the slope of a Ragone plot is the expected discharge time, *t*, of any storage device which initiates a stored energy E_0 and provides a constant power *P* to a certain load. The device is operational only for a limited time *t*. The graph of the maximum available energy, $E(P)$, as a function of P refers to

Table 2.2 Comparison between AC and DC generators [1]

No.	Differentiating property	AC generator	DC generator
1	Definition	AC generator is a mechanical device which converts mechanical energy into AC electrical power	DC generator is a mechanical device which converts mechanical energy into DC electrical power
2	Direction of current	In an AC generator, the electrical current reverses direction periodically	In a DC generator, the electrical current flows only in one direction
3	Basic design	In an AC generator, the coil through which the current flows is fixed while the magnet moves. The construction is simple and costs are less	In a DC generator, the coil through which the current flows rotates in a fixed field. The overall design is very simple but construction is complex due commutators and slip rings
4	Commutators	AC generator does not have commutators	DC generators have commutators to make the current flow in one direction only
5	Rings	AC generators have slip rings	DC generators have split-ring commutators
6	Efficiency of brushes	Since slip rings have a smooth and uninterrupted surface, they do not wear quickly and are highly efficient	Both brushes and commutators of a DC generator wear out quickly and thus are less efficient
7	Short Circuit possibility	As the brushes have high efficiency, a short circuit is very unlikely	Since the brushes and commutators wear out quickly, sparking and short circuit possibility is high
8	Armature	In case of AC generators, the armature is always the rotor	In case of DC generators, the armature may be either rotor or stator
9	Rotating parts	The rotating part in an AC generator is low current high resistivity rotor	The rotating part in a DC generator is generally heavy
10	Current induction	In AC generator, the output current can be either induced in the stator or in the rotor	In DC generator, the output current can only be induced in the rotor
11	Output Voltage	AC generators produce a high voltage which varies in amplitude and time. The output frequency varies (mostly 50–60 Hz)	DC generators produce a low voltage when compared to AC generator which is constant in amplitude and time, i.e., output frequency is zero
12	Maintenance	AC generators require very less maintenance and are highly reliable	DC generators require frequent maintenance and are less reliable
13	Types	AC generators can of varying types like three-phase generators, single-phase generators, synchronous generator, induction generator, etc.	DC generators are mainly two types which are separately excited DC generator and self-excited DC generator. According to field and armature connection, they can be further classified as DC series, shunt, or compound generators, respectively
14	Cost	The initial cost of AC generator is high	The initial cost of DC generator is less when compared to AC generators
15	Distribution and Transmission	The output from AC generators is easy to distribute using a transformer	The output from DC generators is difficult to distribute as transformers cannot be used
16	Efficiency	AC generators are very efficient as the energy losses are less	DC generators are less efficient due to sparking and other losses like copper, eddy current, mechanical, and hysteresis losses
17	Applications	It is used to power for smaller motors and electrical appliances at homes (mixers, vacuum cleaners, etc.)	DC generators power very large electric motors like those needed for subway systems

the Ragone plot. The discharge efficiency of the system can then be estimated via $\eta = \frac{E(P)}{E_0}$.

The holy grail of systems are ones that occupy the upper right corner of Fig. 2.10, which is not surprisingly occupied by internal combustion engines and gas turbines. In both cases of automotive and airplane propulsion, the engines are capable of varying their output to accommodate the demand needs (driving uphill vs. cruising or during takeoff vs. cruising altitude). This versatility has made gas turbines and

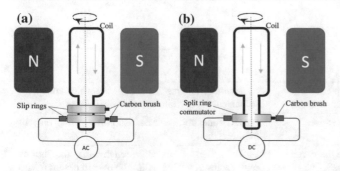

Fig. 2.9 **a** Alternating current generator and **b** direct current generator

diesel engines driving electrical generators the most flexible (yet the most expensive) option whenever peak demand is reached, and extra power is needed for a short period of time.

Whenever a new storage system design is suggested, its performance has to be examined and plotted on the Ragone diagram to determine the range it can cover. Well-established storage systems (lead–acid batteries, fuel cells, etc.) are used as benchmarks to quantify the advantage of a proposed system over existing technologies. The flywheel storage systems are the only one of the mechanical systems discussed in this book that is present in the Ragone plot shown in Fig. 2.9. The location this technology occupies on the Ragone plot is very promising, as it appears to be competing with the most advanced chemical storage systems currently known (the Li–ion batteries). The discussion and analysis of the mechanical systems used for energy storage in this book will cover the expected contribution of each on a Ragone plot. The reader is encouraged to always think in terms of where the storage system in mind fits on the Ragone

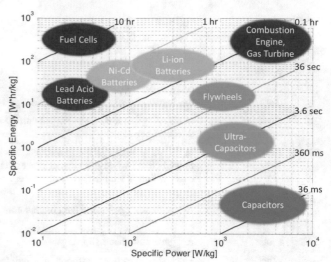

Fig. 2.10 Ragone plot of most common energy storage mechanisms [2]

plot and explores whether and how the system parameters can be tuned to reach the upper right corner of the graph.

References

1. BYJU, Difference Between AC And DC Generator. [Online]. https://byjus.com/physics/difference-between-ac-and-dc-generator/
2. S.J. Moura, J.B. Siegel, D.J. Siegel, H.K. Fathy, A.G. Stefanopoulou, Education on vehicle electrification: battery systems, fuel cells, and hydrogen, in *2010 IEEE Vehicle Power and Propulsion Conference* (2010), pp. 1–6

3.1 Energy Storage in Elastic Components

Elastic elements are among the earliest utilized energy storage techniques in history. Strings in bows and elastic materials in catapults were used to control energy storage and release in ancient war times. The range and momentum of the projectile depended on the mechanical properties of the elastic material launching them. These properties are applied assuming that the material is deformable, thus are different from studying nondeformable (rigid) bodies in statics or in dynamics, where the main governing equations are $\sum F = 0$ and $\sum F = ma$, respectively. Figure 3.1 shows the tension forces arising in an elastic string resulting from pulling an arrow a certain distance, Δx. It is intuitive to note that the larger the distance, the larger the magnitude of the elastic force, and thus more energy is stored.

According to Hooke's law, any material would elastically deform under external loading up to a certain extension limit. The relationship between the applied load and deformation is usually linear and defined as $F = \text{constant} \cdot$ deformation. The only condition for this relationship to hold is not to induce any permanent (plastic) deformation within the material. This means that once the load is removed, the material will return to its original length and ready for another loading cycle. In order to arrive at a general relationship relating the applied force and deformation independent of the geometry and shape of the body, Hooke's law is usually expressed in terms of stress (force/area) and strain (deformation/original length). A plot of an arbitrary stress–strain diagram is shown in Fig. 3.2, highlighting the main stress states such as the yield point and fracture point. This diagram is generally obtained when a ductile nonferrous specimen is deformed under axial tension or compression. The slope of the elastic region before yield takes place is usually linear which means it is constant (see inset) and is called the modulus of elasticity or Young's modulus. Its values are tabulated for most known materials in the Mechanics of Materials textbooks. For the straight-line part,

Hooke's law is written as $\sigma = E\epsilon$, and it only applies before yield takes place.

The careful examination of the units pertinent to stress and strain indicates that the area under the stress–strain diagram is in fact that of energy per unit volume:

$$Stress = \frac{Force}{Area} \ (\text{N/m}^2) \tag{3.1}$$

$$Strain = \frac{Deformation}{Original\ dimension} \ (\text{m/m}) \tag{3.2}$$

The area under the curve is the multiplication of the stress and strain, and the units are seen to be $\left(\frac{\text{N m}}{\text{m}^2\,\text{m}}\right)$ or $\left(\frac{\text{J}}{\text{m}^3}\right)$. It is also straightforward to note that the area under the elastic region before yielding is reached is the area of the triangle seen in the inset of Fig. 3.2, and is equal to

$$Energy = \frac{1}{2}\sigma \cdot \epsilon \ \ \left(\frac{\text{J}}{\text{m}^3}\right) \tag{3.3}$$

This is the energy that a unit volume of the material can store without being permanently deformed. And since the stress cannot be easily estimated while the material deformation can be easily measured, Hooke's law can be used to rewrite the energy equation to be

$$Energy = \frac{1}{2}E \cdot \epsilon^2 \ \ \left(\frac{\text{J}}{\text{m}^3}\right) \tag{3.4}$$

This relation is analogous to the energy equation for linear springs, where the modulus of elasticity replaces the spring constant, k, and the strain replaces the displacement.

3.2 Linear Springs

The energy stored in linear springs is proportional to the square of the distance, Δx, displaced away (extension or compression) from a certain reference point or datum, as

© Springer Nature Switzerland AG 2020
A. H. Alami, *Mechanical Energy Storage for Renewable and Sustainable Energy Resources*,
Advances in Science, Technology & Innovation, https://doi.org/10.1007/978-3-030-33788-9_3

(a)	**(b)**

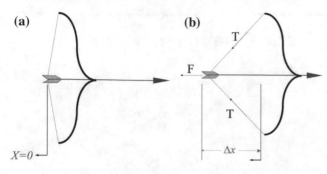

Fig. 3.1 The elastic force arising from pulling a bowstring a distance Δx with an equal and opposite force, F

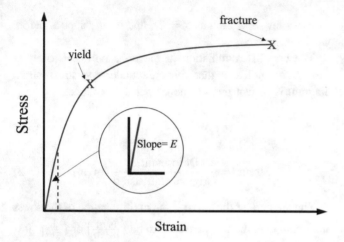

Fig. 3.2 Stress–strain diagram of a nonferrous elastic material

shown in Fig. 3.3. Similar to elastic elements, the spring force is defined as

$$F_s = k\Delta x \tag{3.5}$$

where k is the spring constant, analogous to the Young's modulus in Hooke's relation.

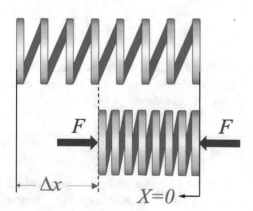

Fig. 3.3 Free-body diagram of a linear spring

Since the spring is not expected to be loaded beyond its elastic limit, the relationship of Eq. (3.5) is a linear one, and the energy stored is the area under that triangular curve:

$$E = \frac{1}{2}k\Delta x^2 \text{ (J)} \tag{3.6}$$

Based on the expected energy level to be handled, a linear spring with a certain constant, k, should be selected. The larger the k, the larger the slope (see the inset of Fig. 3.2) and the higher the spring stiffness.

3.3 Torsional Springs

Torsional springs as energy storage devices are used in simple mechanical devices, such as timekeeping pieces and mousetraps among others. The analogy of force and displacement holds as for other elastic elements, but for torsional springs the displacement is measured in terms of rotation angles, θ (rad), and the applied forces as a torque, T (N m). The governing equation is then written as

$$T_s = k_G\Delta\theta \tag{3.7}$$

The torsional spring constant, k_G, is also a material property and the material selection happens according to the expected torsional loads, or the maximum allowable angular translation. A torsional spring is depicted in Fig. 3.4.

The energy stored within a torsional spring is calculated in a similar manner to their linear counterparts, noting that the slope of the torque (load) versus deflection (angle) is the torsional spring constant, and the triangular area under the curve is also the energy stored in the spring:

$$E = \frac{1}{2}k_G\Delta\theta^2 \text{ (J)} \tag{3.8}$$

Old winding alarm clocks used such springs, mostly coupled with ratchets to control the energy release and allow the clock to operate for days between rewinds.

3.4 Hyperelastic Material

Nonlinear elastic materials are ones that do not obey Hooke's law which correlates the load/displacement in a linear fashion. They respond with pure elasticity to excessive amounts of load, and their strain levels can sometimes go beyond 100% (sometimes up to 700%) without causing failure. Another property that is important in formulating their stress–strain behavior is their incompressible nature, which means that regardless of shape changes, the overall volume remains constant. The stress–strain behavior of such materials under uniaxial tension is shown in Fig. 3.5.

Fig. 3.4 Torsional springs varieties. Note that in (b) A is the arbor diameter and OD is the outside diameter in the free condition [1]

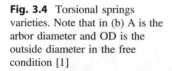

(a)

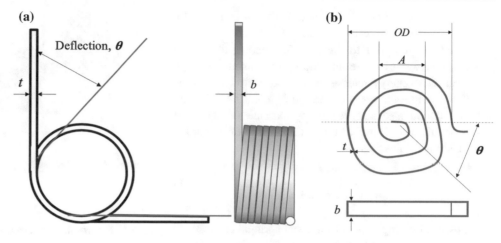

(b)

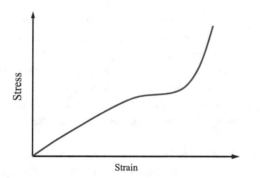

Fig. 3.5 Stress–strain diagram of a hyperelastic material under uniaxial tension

There are various models that are used to approximate the behavior of such materials under loading, most notable of which is the Mooney–Rivlin model which if formulated around the strain energy density function that uses empirically derived coefficients from uniaxial tension experiments to approximate the material behavior. Once the stress–strain relationship is established, the area under the curve can be calculated to establish the energy that can be absorbed in the material without rupturing (failure).

Hyperelastic materials include most polymers and rubbers, which are materials normally used to absorb energy for vibration isolation applications in cars and machinery. With the large area under the stress–strain curve (energy per unit volume) coupled with elasticities that extend well beyond the 100% strain, they are good candidates to replace linear and torsional springs, as long as the application does not involve extreme temperatures or corrosive conditions that might jeopardize the integrity of the material. More research is still missing on the applications of such materials for the macro-applications energy storage, rather than the micro-level research already available in literature, that helps derive the expected behavior of such materials in numerical models.

3.5 Energy Storage Applications

The applications of elastic elements are energy conversions into either potential or kinetic energies. For example, the potential energy stored within a coiled torsional spring is enough to operate a timekeeping machine (clock) for a certain period of time depending on the coil dimensions and material. The winding of a torsional spring increases the torque that is a result of the material elastic resistance, and once the spring is wound, its uncoiling (unwinding) motion is controlled by a system of gears, ratchets, and sometimes cams to speed up or slow down the motion. Proper timing is kept via the mechanical design of the gears (number of teeth, gear diameter, etc.).

An equation that can be used to estimate the torque (N m) as a function of the winding of the spring (angular deflection, θ, in revolutions) is as follows [2, 3]:

$$T = \frac{\pi E b t^3}{6L} \theta \qquad (3.9)$$

where E is the modulus of elasticity (MPa), L is the length of the coil (mm), b is the coil width (mm), and t is the thickness of the coil (see Fig. 3.4b). Once the torque is found, the maximum stress that can be imposed on the spring is then estimated from the following equation:

$$\sigma = \frac{6T}{bt^2} \qquad (3.10)$$

This stress value is compared with the mechanical properties of the proposed candidate materials to ensure that the spring will handle the operation while remaining elastic, and also deduce the useful endurance limit (fatigue life) to determine the period of service. Other important design parameters that do not appear in the torque or the stress equations are (i) the outside diameter in the free condition

(*OD*) and (ii) the arbor diameter (*A*), shown in Fig. 3.4b. The spring would have to be compressed and thus suffer unwanted stresses prior to the intended loading due to the mismatch between its size and the space it should occupy. This can cause energy release issues and/or decrease in the useful life before the onset of fatigue failure. The following formula can be used in order to approximate the minimum space required to host a torsional spring of a certain dimension and thickness:

$$OD = \frac{2L}{\pi\left(\frac{\sqrt{A^2 + 1.27Lt} - A}{2t} - \theta\right)} - A \qquad (3.11)$$

Thus, the limiting cases of torsional springs' applications would be geometrical (the minimum winding diameter given a certain thickness and length of the spring), the yield strength of the material, as well as its fatigue life.

As a simple design example, assume that a spiral torsional spring is made of 0.81-mm-thick by 6.35-mm-wide 1070 carbon steel and must deliver a moment of 508 N m at 135° deflection. The spring is to work over a 6.35-mm-diameter arbor. What is the active length of the material required, the stress imposed on the spring, and the spring outside diameter in the free condition (OD)?

It is quite straightforward to find the length by transposing the torque formula of Eq. (3.9) to become

$$L = \frac{\pi E b t^3}{6T} \theta$$

where $\theta = \frac{135°}{360°} = 0.375$ revolutions and E for 1070 carbon steel 201 GPa. By plugging the numbers into the equation above, we find $L = 272.3$ mm. This length does not include any special configurations for fixing the ends of the spring.

The expected stress to be imposed on the spring is calculated from Eq. (3.10):

$$\sigma = \frac{6T}{bt^2} = \frac{6(508)}{(6.35)(0.81)^2} = 724 \, \text{MPa}$$

The OD can be readily calculated from Eq. (3.11):

$$OD = \frac{2L}{\pi\left(\frac{\sqrt{A^2 + 1.27Lt} - A}{2t} - \theta\right)} - A$$

$$= \frac{2(272.3)}{\pi\left(\frac{\sqrt{(6.35)^2 + 1.27(272.3)(0.81)} - 6.3}{2(0.81)} - 0.375\right)} - 6.35$$

$$= 19.35 \, \text{mm}$$

Another application of elastic elements is the incorporation of linear springs in automatic and semi-automatic pistols (see Fig. 3.6). According to Newton's third law of motion that links any action to an equal and opposite reaction, once a pistol is fired, a large force is imparted on the projectile

Fig. 3.6 Firearm showing the recoil spring in red [4]

causing it to escape the muzzle at a high velocity. An equal and opposite force will in turn affect the pistol and will knock it out of the hand of its operator, which would result in injuries and inaccurate shots. A recoil spring is installed within the pistol to not only absorb the force but also store it for ejecting the spent casing and to load the next cartridge and returning the pistol to its original, ready-for-firing position.

The spring has to be chosen so as not to have too high a spring constant value which will not allow the spring to cycle all the way back and produce the desired effect. It also should not have too low of a value for the spring constant that will cause the spring to compress too easily and thus it will introduce a delay in the operation of the pistol by returning to the firing position at a slower rate and thus causing follow-up shots to take significantly longer time.

Since the rate of change of momentum of any object is the resultant applied force on it, remember that the momentum, *p*, can be defined as $p = mv$, where v is the velocity of the particle and the rate of change of velocity will be the acceleration $\left(\frac{dp}{dt} = ma = F\right)$.

Thus, the spring selection for energy storage during the discharge of a firearm to release the spent shell and load the next one has to be done on the basis of the following equations:

$$E = \frac{1}{2}mv^2 = \frac{p^2}{2m} \qquad (3.12)$$

This energy is imparted on the projectile to cause it to move forward to the farthest distance. The exploding propellant expands in the confines of the firearm and creates a large hydrostatic pressure that causes the projectile to exit the firearm at a high velocity, *v*. This will cause the firearm to recoil with the same force as the bullet leaving the muzzle. The average weight of a bullet is ~10 g, and the muzzle velocity is ~850 m/s (muzzle velocities range from approximately 120 to 370 m/s in black powder muskets, to more than 1,200 m/s in modern rifles with high-performance cartridges). Thus, the energy that needs to be stored within the spring (area under the force–displacement) is around 3.6 kJ. A spring of that has 1 mm wire thickness, coiled in 50 loops with a length of 50 and 2 cm loop diameter made from steel would have to have a spring constant of ~ 20 N/m to handle such energy. The spring design procedure is lengthy but systematic and can be found in various specialized textbooks and often online calculators to streamline the spring selection process.

References

1. L.P. Pook, An introduction to coiled springs (mainsprings) as a power source. Int. J. Fatigue **33**(8), 1017–1024 (2011)
2. F. Rossi, B. Castellani, A. Nicolini, Benefits and challenges of mechanical spring systems for energy storage applications. Energy Procedia **82**, 805–810 (2015)
3. Springs and Things Incorporated, Spiral Torsion Springs. [Online]. http://www.springsandthings.com/pdf/spiral-torsion-springs.pdf
4. S. Hoober, What You Need To Know About Recoil Springs (2017). [Online]. https://gunbelts.com/blog/handgun-recoil-springs/

Thermal Storage

4.1 Thermal Storage

The thermal storage techniques can be classified in a variety of ways based on the energy source and the intended application. The main classification of thermal energy storage (TES) systems stems from whether or not phase transformation of the material is to be utilized for storage. If phase transformations are permitted, it is known to allow the system to hold on to the heat for longer times. This is known as latent heat storage that uses phase-change materials (PCMs), or materials that can change phase (from solid to liquid or liquid to vapor). This is different than the straightforward thermal storage that occurs without the phase change, which is known as sensible heat storage, that usually happens in well-insulated vessels and containers. An example of such a device is shown in Fig. 4.1.

When it comes to working fluids, water is by far the best thermal storage medium. It has the highest specific heat value at 4.182 kJ/kg, as well as other operational benefits, such as its availability and recyclability from one phase to another with benign effects on the carrying conduits (unless it is contaminated with pH altering substances that can cause unwanted corrosion or erosion). Table 4.1 lists some common materials with the nominal values of specific heat, c_p, thermal conductivity, k, and diffusivity. It should be noted that the thermal diffusivity, α, of a material is an indication of how fast heat can be transferred through it versus how much of this heat is stored within the material $\left(\alpha = \frac{k}{\rho c_p}\right)$ and thus the lower it is, the better the material is for storage.

Sensible heat storage is straightforward. The material with the lowest thermal diffusivity would be most suitable, and the phase chosen (solid, liquid, or gas) depends on the application and available installation. It is interesting to note that in ancient times people would expose bricks to the sun all day long, and then place them at strategic locations (where air circulates) in their dwellings to provide heating for the latter part of the day and into the evening. The concept is still being used in England and Australia (called Storage Heater) but for load shifting purposes. This system utilizes ceramic bricks placed on an electric heater (charging) during times where the electric tariff is cheaper for low demand cycles. Then, once the peak demand occurs with higher cost per kWh, the heater is shut off and the bricks would release the heat (discharge) to the surrounding by radiation, convection, or conduction. Special fans would also be used to enhance the forced convection component of the heat transfer. The bricks are one of the more attractive solid media for this purpose as it has one of the lowest thermal diffusivity values (Table 4.1). Bricks also would not incinerate and be a fire hazard compared with the more thermally attractive wood or break in the case of using glass, which also has an attractive thermal diffusivity value.

The sensible heat equation is quite simple, and by recalling the discussion in Chap. 2, the governing equation is the first law of thermodynamics for heat addition in the absence of any shaft work:

$$Q = mc_p\Delta T \tag{4.1}$$

The straightforward option is for the material with the highest specific heat capacity to be used to ensure sufficient capacity is available to store the heat load during system charging. Alternatively, the utilization of the proper quantity (mass) of a material with moderate heat capacity is required to handle the load. And since the heat in this case is expected *not* to be easily transferred, a layer of insulation (low thermal conductivity with low specific heat) should be applied to the storage reservoir as shown in Fig. 4.1 to prevent losses into the surrounding environment. The same goes for any piping or fittings leading to or from the supply/supplied device to ensure the highest operational efficiency.

4.2 Latent Energy Storage

Changing the phase of the heat storage material means that the energy content of this material is increasing without changing its temperature. Thus, a phase-change material can be utilized within the system, and its latent energy can be

© Springer Nature Switzerland AG 2020
A. H. Alami, *Mechanical Energy Storage for Renewable and Sustainable Energy Resources*,
Advances in Science, Technology & Innovation, https://doi.org/10.1007/978-3-030-33788-9_4

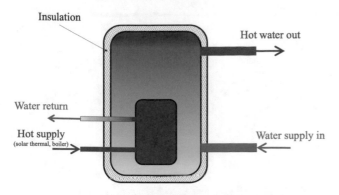

Fig. 4.1 The components of a sensible heat storage system

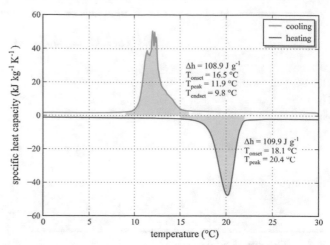

Fig. 4.2 DSC measurement for the eutectic salt hydrate mixture $Zn(NO_3)_2 \cdot 6H_2O$, $Mn(NO_3)_2 \cdot 4H_2O$, and KNO_3 [3]

given to (charge) or extracted from (discharge) the heat transfer medium to achieve this goal. In this case, the material can be originally a solid changing phase into a liquid, or a liquid changing phase into a gas. Figure 2.5 shows how the energy content of a single phase changes almost linearly when heat is added that causes a temperature increase in the heat transfer medium up to the onset of the phase change that happens isothermally.

The equation for the heat transfer that includes the phase-change material is a modification of Eq. (4.1) as follows:

$$Q = mc_p\Delta T + mh_{fg} \qquad (4.2)$$

where h_{fg} is the latent heat of fusion (for transformation from solid to liquid) or latent heat of evaporation (for transformations from liquid to solid) in J/kg. Note that there is no pertinent temperature change for the latent part of this equation as the phase transformation takes place at constant temperature ($\Delta T = 0$). Of course, this assumption that the latent heat is added at constant temperature is an idealized

one, as the "thermal inertia" of the material does not allow the instantaneous heat addition to take place. This is a positive thing in a latent thermal storage system as it allows abundant input heat to first heat the material up (sensible), change the phase of the material (latent), and finally heat up the second phase of the material (sensible again).

Figure 4.2 shows a differential scanning calorimetry test of a salt material, with the peaks showing the latent energy intake/release signifying a phase change. Note that the phase change happens over a temperature range, rather than instantaneously at one temperature value, which is beneficial for the storage purpose. Also, notice how the latent energy peaks occur at different temperature values rather than at the same temperature when heating or cooling, which also indicates the hysteresis effect that the material suffers from during adding/removing heat.

Table 4.1 Thermal properties of common materials [4]

Material	Density ($kg\,m^{-3}$)	Specific heat ($J\,kg^{-1}\,K^{-1}$)	Thermal conductivity ($J\,m^{-1}\,K^{-1}$)	Thermal diffusivity ($10^6\,m^2s^{-1}$)
Clay	1,458	879	1.4	1.10
Brick	1,800	837	1.3	0.86
Sandstone	2,200	712	1.7	1.08
Wood	700	2,390	0.17	0.10
Concrete	2,000	880	1.8	1.02
Glass	2,710	837	1.05	0.46
Aluminum	2,710	896	237	97.60
Steel	7,840	465	54	14.81
Air	1.27	1006	0.024	18.8
Water	988	4,182	0.6	0.14

4.2.1 Available Materials for Latent Heat Storage: Inorganic Materials

There are several candidate materials that can be suitable for latent heat storage applications. They can be classified mainly as either inorganic or organic materials. The former is in the form of salts and hydrated salts usually used for high-temperature applications. The phase transformation here is not limited to solid to liquid and vice versa, but also includes solid-state eutectic reactions. Some pertinent latent heat data is available for some inorganic materials and is shown in Table 4.2 where phase (solid to liquid) transformations occur, while Table 4.3 present similar data for inorganic materials where the solid-state phase transformations take place and is capable of temporary storage of heat. In this case, it is important to quote the accompanying changes in entropy at the transition temperature in order to be able to quantify the thermal capacity of the phase under consideration.

In most solid-state transformation instances, the general crystal structure of the material at or around the solid-state transformation temperature is not very different between the starting and end phases. Thus, the transition entropy values are not expected to be high, and most of the heat added to the material causing entropy change is due to the phase transformation from solid to liquid rather than the solid-state one. This explains the higher value of the melting entropy for a material such as FeS compared with its transition entropy. On the other hand, there are materials where the transition entropy is high due to unusual vibrational amplitude (inter-site mobility) of some atomic species; thus, materials such as AgI have a high mobility and this is reflected at the large value of transition entropy.

4.2.2 Organic Phase-Change Materials

There are several organic materials available for latent energy storage, and the data on the melting temperatures and heats of fusion for organic, fatty acids, and aromatic materials are given in the tables from Tables 4.4, 4.5 and 4.6.

4.3 Experimental Latent Heat Trial

A proof-of-concept setup is available at the Energy Storage Laboratory at the University of Sharjah to assist in exploring the benefits of using a phase change material as latent heat storage medium. The experimental setup, shown in Fig. 4.3, consists of two identical 25-cm-diameter copper cylinders with a wall thickness of 1.5 mm. A smaller cylinder filled with water is internally fastened to the bottom of one cylinder and the space between them is filled with paraffin wax. The second cylinder is also filled with the same amount of water as the inner cylinder. This means that both cylinders receive the same amount of solar radiation as both are identical in size, material, and surface finish. Both cylinders are also equipped with K-type thermocouples to measure the temperature of the water, and readings are taken by a data acquisition (DAQ) setup connected to a PC.

During the charging cycle that started at 10:00 am, the system of two cylinders is left flat under the sun, where both cylinders are expected to receive the same amount of incident solar radiation. The temperature of the water from the first cylinder is noticed to rise at a higher rate than the inner one surrounded by the paraffin wax, as shown in Fig. 4.4. The charging cycle is terminated at around 2:00 pm as the system is pulled into a shaded area. The paraffin wax in the first cylinder is completely in liquid form by that time.

During the discharge cycle, the temperature of the system with water decreases rapidly, while the one containing paraffin wax is seen to decrease at a lower rate until there is a point where the temperature profile for the paraffin wax cylinder is higher than its water-only counterpart. This is where the effective thermal storage is achieved, as shown in the shaded area of Fig. 4.4.

The background grid line in Fig. 4.4 is left on intentionally. Since the water quantity (mass) in both setups is the

Table 4.2 Latent heat values for inorganic materials [4]

Phase	Melting point (°C)	Heat of fusion $(MJ\,kg^{-1})$
NH_4NO_3	170	0.12
$NaNO_3$	307	0.13
NaOH	318	0.15
$Ca(NO_3)_2$	561	0.12
LiCl	614	0.31
$FeCl_2$	670	0.34
$MgCl_2$	708	0.45
KCl	776	0.34
NaCl	801	0.50

Table 4.3 Solid-state phase transformation and melting entropy data for some materials [4]

Material	Transition temperature (°C)	Melting temperature (°C)	Transition entropy (J.mol⁻¹.K⁻¹)	Melting entropy (J mol⁻¹ K⁻¹)
FeS	138	1,190	4.05	21.51
AgI	148	558	14.61	11.33
Ag$_2$S	177	837	8.86	7.01
Na$_2$SO$_4$	247	884	12.5	18.2
Ag$_2$SO$_4$	427	660	26.66	19.19
Li$_2$SO$_4$	577	860	29.2	7.9
LiNaSO$_4$	518	615	31.2	Small

Table 4.4 Latent heat values for organic PCMs [4]

Material	Melting point (°C)	Heat of fusion (k.J.kg⁻¹)
Paraffin wax	64	173.6
Polyglycol E400	8	99.6
Polyglycol E600	22	127.2
Polyglycol E6000	66	190.0

Table 4.5 Latent heat values for fatty acids used as PCMs [4]

Material	Melting point (°C)	Heat of fusion (k.J.kg⁻¹)
Stearic acid	69	202.5
Palmitic acid	64	185.4
Capric acid	32	152.7
Caprylic acid	16	148.5

Table 4.6 Latent heat values for aromatic materials used as PCMs [4]

Material	Melting point (°C)	Heat of fusion (k.J.kg⁻¹)
Biphenyl	71	19.2
Naphthalene	80	147.7

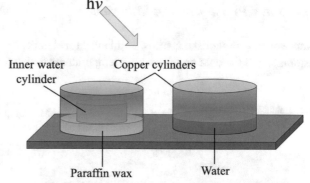

Fig. 4.3 Setup for the latent heat storage experiment

same, then the areas under the red (with wax) curve versus the area under the blue (no wax) curve would give a quick indication of the percent increase in the thermal energy storage effectiveness for the (with wax) system over the (no wax) one. In this case, the thermal storage capacity of the system has increased by around 20%. Of course, there are a few sources of losses in the current setup (no thermal insulation around the cylinders, the system is placed in a

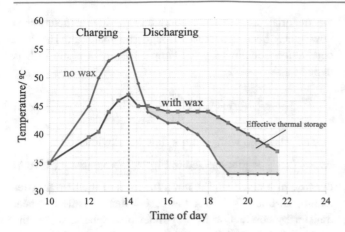

Fig. 4.4 Temperature profile of the latent heat storage experiment

glass frame to decrease convective heat loss but giving rise to reflective losses in the received solar irradiation, etc.) but since both cylinders are indistinguishable and under identical operating conditions, a sufficiently accurate estimate of storage gains achieved via can be made.

4.4 Thermal Storage for Solar Thermal Power Plants

To put the role PCM materials play in power cycles into perspective, it is useful to examine the "compulsory" provisions for latent energy storage that is part of solar thermal power plants that utilize concentrated solar power (CSP). Figure 4.5 shows the coupling between such power plant and a classical Rankine power cycle, complete with a steam

turbine that generates the mechanical power needed to drive an electrical generator.

Since water has the highest specific heat, it would be infeasible to use it as the working fluid in a CSP cycle directly. Water will take too long to heat up and will require high pressures to keep water from evaporating at around 100 °C (the higher the pressure, the more energy can be packed into the water without losing it to evaporation, just like in a pressure cooker). There is also the transient nature of incoming solar radiation that necessitates the rapid conversion of the incident energy into sensible heat that will be eventually absorbed by the working fluid. Thus, using a salt (usually a calcium–potassium–sodium–nitrate-based mixture) that has a low melting point, a low value of heat of fusion is widely implemented. This working fluid is passed through the CSP array where it changes phase (becomes molten) and flows into a reservoir that acts as a buffer to ensure no shortage of salt supply will be suffered in the next stage, which is the main heat exchanger that transfers the heat into water. This heat exchanger is unmixed, meaning that the salt and water have their independent conduits and will not come in physical contact with one another, only thermal contact. The same amount of heat is transferred between the two fluids, but since the equilibrium energy equation is $\dot{Q} = \dot{m}c_{p,\text{water}}\Delta T = \dot{m}c_{p,\text{salt}}\Delta T$, the temperature rise in the water is expected to be much less than that in the salt since the specific heat of water is much greater than that of the salt (water has a little less than four times higher specific heat). But at high pressures (operating pressures on the waterside reach 10–15 bar), the thermal energy can be stored in the steam line until it is expanded in the steam turbine, where the Rankine cycle is also closed at the condenser, followed by the pump.

Fig. 4.5 Depiction of a concentrated solar power plant with a Rankine power cycle

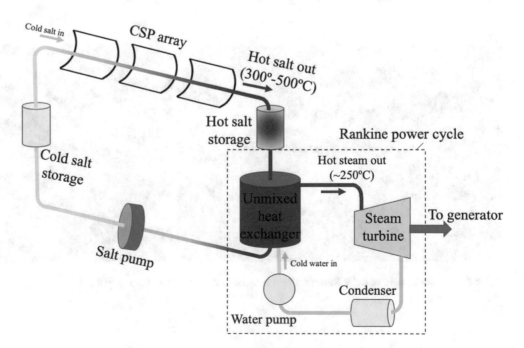

The molten salt is then pumped back into the cold reservoir before completing its respective cycle and passing once more through the CSP array. It should be noted that the storage tanks for the molten salt also contain electrical heaters to ensure the salt remains at its pumpable molten state in the absence of thermal energy from the sun (at night or during overcast conditions). The thermal insulation of these tanks is also very important to prevent any heat loss to the environment in a manner similar to that shown in Fig. 4.1.

It is worth noting that the molten salt cycle is applicable also in the case of a heliostat (central receiver with a field of mirror arrays) system instead of the CSP array. Since this is a point-receiver system, as opposed to CSP being a line concentrator, the expected operational temperatures of the salt side are higher and can range in 400–600 °C for the molten salt side. This requires the waterside to operate at higher pressures to accommodate the expected higher temperatures that can range 300–450 °C. Such a system is shown in Fig. 4.6 operated by Solar Reserve [1] where salt storage tanks can be seen integrated around the central receiver installation to handle the molten salt cycle.

The following example will emphasize the importance of using heat transfer relations for steady-state conduction across the molten salt tank. Since the tank can be considered as a large reservoir that is always full of salt, heat transfer problems can be approximated to operate at steady-state conduction conditions. Thus, the thermal resistance circuit approximation can be applied, making the sizing of the reservoir an easy task. Remember that heat, just like electrical current, passes whenever there is a temperature difference between two points (analogous to a voltage difference). There are geometrical parameters, i.e., the surface area, A, length, $\Delta x = L$ as well as parameters related to the material, i.e., thermal conductivity, k, that determines how much heat will pass through the material. The heat conduction equation for the simple case of a plane wall is defined as

$$\dot{Q} = -kA\frac{dT}{dx} \approx -kA\frac{\Delta T}{L} \qquad (4.3)$$

The analogy with Ohm's law is obvious: $\Delta V = \frac{I}{R} \equiv \Delta T = \frac{Q}{\frac{L}{kA}} = \frac{Q}{R}$ as summarized in Fig. 4.7. Note that the area in the case of a plane wall is not a function of the heat transfer length, L, but this is not the case when considering heat transfer by conduction in a cylinder or a sphere, where the area is a function of the radial heat transfer since the surface area is $2\pi r L$ and $4\pi r^2$, respectively.

The example at hand examines a spherical container of inner radius $r_1 = 40$ cm, outer radius $r_2 = 41$ cm, and thermal conductivity $k = 1.5$ W/m·°C that is used to store molten salt and to keep it at 300 °C at all times as part of a CSP plant as shown in Fig. 4.8. To ensure the salt remains molten, the outer surface of the container is wrapped with a 1500-W electric strip heater and then insulated. The temperature of the inner surface of the container is observed to be nearly 300 °C at all times. Assuming 10% of the heat generated in the heater is lost through the insulation, the following is required:

(a) express the differential equation and the boundary conditions for steady one-dimensional heat conduction through the container,
(b) obtain a relation for the variation of temperature in the container material by solving the differential equation, and

(a)   **(b)**

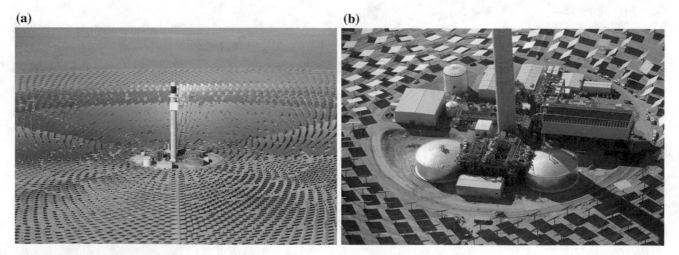

Fig. 4.6 **a** Heliostat field of Solar Reserve where **b** molten salt storage tanks are shown [1]

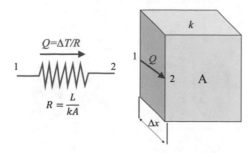

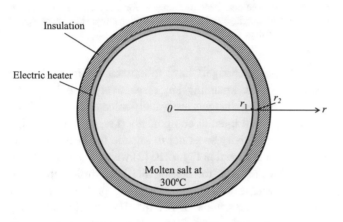

Fig. 4.7 Electrical circuit analogy for steady-state convection through a plane wall

Fig. 4.8 Spherical container for storing molten salt

(c) evaluate the outer surface temperature of the container. Also, determine how much salt at 300 °C this tank can supply steadily if the cold salt enters at 150 °C.

To solve this problem, we first note that only 90% of the strip heater power is transferred to the container, and we can determine the heat flux (heat transfer per unit surface area) through the outer surface by the following formula:

$$\dot{q} = \frac{\dot{Q}}{A_2} = \frac{\dot{Q}}{4\pi r_2^2} = \frac{0.9 \times 1500}{4\pi (0.41)^2} = 639 \ W/m^2$$

Assuming that the heat transfer is one-dimensional and in the radial r-direction (see Fig. 4.8), the mathematical formulation is expressed as

$$\frac{d}{dr}\left(r^2 \frac{dT}{dr}\right) = 0$$

With boundary conditions, $T(r_1) = T_1 = 300°$ and $k\left(\frac{dT(r_2)}{dr}\right) = \dot{q}$

The differential equation is integrated once with respect to r to yield: $\left(r^2 \frac{dT}{dr}\right) = C_1$

Dividing both sides with r^2 gives

$$\frac{dT}{dr} = \frac{C_1}{r^2}$$

then integrating again

$$T(r) = -\frac{C_1}{r} + C_2$$

Note that C_1 and C_2 are arbitrary constants. Applying the boundary conditions:

@ $r = r_2$:

$$k\left(\frac{C_1}{r_2^2}\right) = \dot{q} \rightarrow C_1 = \frac{\dot{q}r_2^2}{k}$$

@ $r = r_1$:

$$T(r_1) = T_1 = -\frac{C_1}{r_1} + C_2 \rightarrow C_2 = T_1 + \frac{C_1}{r_1} = \frac{\dot{q}r_2^2}{kr_1}$$

Substituting C_1 and C_2 into the general solution,

$$T(r) = -\frac{C_1}{r_1} + C_2 = -\frac{C_1}{r} + T_1 + \frac{C_1}{r_1} = T_1 + \left(\frac{1}{r_1} - \frac{1}{r}\right)\frac{\dot{q}r_2^2}{k}$$

$$= 300 + \left(\frac{1}{0.4} - \frac{1}{r}\right)\frac{639(0.41)^2}{1.5}$$

$$T(r) = 300 + 174.7\left(2.5 - \frac{1}{r}\right)$$

Thus, the outer surface temperature at $r_2 = 0.41$ m is $300 + 174.7\left(2.5 - \frac{1}{0.41}\right) = 310$ °C.

It is now straightforward to calculate the maximum rate of supply for the salt by using

$$\dot{Q} = \dot{m}c_p \Delta T$$

Knowing that the heater supplies heat at a rate of $1500 \times 0.9 = 1350$ W, the temperature difference is 150 °C and the specific heat of the salt is 1.3 kJ/kg.°C: the system can supply up to 0.005 kg/s (or 19 kg/h) of the molten salt at 300 °C.

4.5 Thermal Storage for Compressed-Air Energy Storage (CAES) Systems

The compressed-air energy storage systems will be discussed in full detail in Chap. 7. This section deals only with storing the heat generated due to the compression of air to utilize it later when the system is discharging, and the air is decompressing. The low temperatures that accompany the expanding air can cause the moisture content to either liquify or freeze, which has detrimental effects on the subsequent air turbine that is used to run the electrical generator.

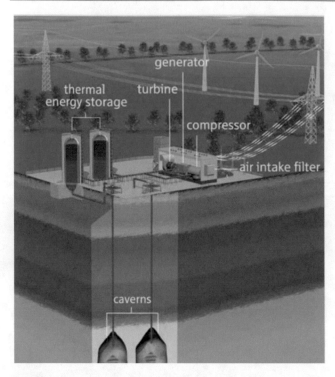

Fig. 4.9 ADELE compressed-air energy storage (CAES) system showing the heat exchangers for thermal energy storage

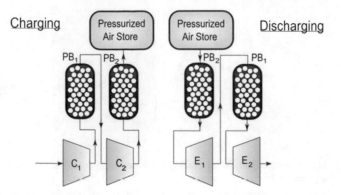

Fig. 4.10 Packed-bed heat exchangers for heat storage during air compression (charging) and heat supply during expansion (discharging) in a CAES system [2]

Such heat exchangers have to extract/supply the heat to the compressed/expanding air, respectively at a high rate. Thus, the heat exchangers are usually selected to have a high surface area and internal components (i.e., pipes) with high thermal conductivity so as not to impede the heat flow. One such system is shown in Fig. 4.10 [2] where packed-bed heat exchangers are used for storing the heat during the compression cycle at a maximum operating pressure of 80 bar and an air storage volume of 182 m^2. The reported efficiency of such a system is over 70%, which means that external heaters may be used but are not necessary to compensate for the heat loss to the environment.

The heat storage in such cases takes place in a container similar in concept with the one shown in Fig. 4.1, but for the purpose of the rapid flow of air during charging and/or discharging, the heat exchangers will also act as thermal storage whenever the need arise. The process uses the same heat generated during compression (note the relationship between the temperature and pressure: $PV = \rho RT$) and supplies it back to the stored air during decompression. This is analogous to checking in your coat at a ball for storage, and later gets it back when it is time to leave. The heat exchangers fit at the inlet of the underground storage caverns where air is stored (heat extraction) and the outlet of the caverns (heat recovery) as shown in ADELE installation of Fig. 4.9, which will be discussed in detail later.

References

1. Solar Reserve, Leading the world in solar power with storage (2018). https://www.solarreserve.com/
2. E. Barbour, D. Mignard, Y. Ding, Y. Li, Adiabatic compressed air energy storage with packed bed thermal energy storage. Appl. Energy **155**, 804–815 (2015)
3. A. Efimova, S. Pinnau, M. Mischke, C. Breitkopf, M. Ruck, P. Schmidt, Development of salt hydrate eutectics as latent heat storage for air conditioning and cooling. Thermochim. Acta **575**, 276–278 (2014)
4. R.A. Huggins, *Energy Storage* (Springer, 2010)

5.1 Flywheel Storage Systems

The first known utilization of flywheels specifically for energy storage applications was to homogenize the energy supplied to a potter wheel. Since a potter requires the involvement of both hands into the axisymmetric task of shaping clay as it rotated, the intermittent jolts by the potter foot meant that the energy supply would have short-term peaks when a pulse of energy is supplied, then it will trail off during the time it takes the potter foot to reach the wheel again. The friction from the bearings of the setup and the applied torque from the potter hands will rapidly consume the energy supplied by the initial kick. The addition of a flywheel is expected to assist in the stabilization of the operation of the device. The flywheel in fact is simply just an extra mass that will keep the kinetic energy of the system, defined as $\frac{1}{2}I\omega^2$, as steady as possible by normalizing the discrepancy of energy charge/discharge levels. In this particular case, this is achieved through increasing the physical inertia of the system (the resistance of the system to any change of its motion). Figure 5.1 shows examples of the progression of flywheel applications through time and different technologies. Note that the common factor of utilizing a flywheel for energy storage is the reciprocating nature of the energy supply, where the input energy peaks and then diminishes. Most intermittent or reciprocating prime movers that are characterized by providing a power stroke as part of their operational cycle need a flywheel to store excess energy during input peaks that would prove useful during the return/exhaust stroke for the normalization of the chrono-energetic exchange cycle of the system.

Just like the mass, m, in the linear kinetic energy equation $\left(\frac{1}{2}mv^2\right)$, the moment of inertia, I, in the rotational kinetic energy definition $\left(\frac{1}{2}I\omega^2\right)$, represents the resistance of the moving body to changes in its momentum. This term increases with the increase of the mass of the wheel as well as with the increase in its radius, r, around the axis of rotation (remember that $I = \frac{1}{2}mr^2$). Thus, the bigger the radius of a rotating body, the more difficult it would be to get it to rotate but at the same time it would be difficult to slow it down or instantaneously bring it to a complete stop. In this inertia/applied force, interaction lays the basic concept behind the utilization of flywheels for energy storage in any mechanism. Some common values of the moment of inertia are given in Fig. 5.2 for different shapes and configurations.

Table 5.1 gives examples of flywheel characteristic based on the technology available.

The flywheel storage technology is best suited for applications where the discharge times are between 10 s to two minutes. With the obvious discharge limitations of other electrochemical storage technologies, such as traditional capacitors (and even supercapacitors) and batteries, the former providing solely high power density and discharge times around 1 s and the latter providing solely high energy density discharge times of over 100 s, flywheels have significant advantages for discharge times between 1 and 100 s and discharge powers above 20 kW as shown in Fig. 5.3.

The components of a flywheel energy storage systems are shown schematically in Fig. 5.4. The main component is a rotating mass that is held via magnetic bearings and enclosed in a housing. The magnetic bearings have the same polarity as the rotor itself and thus generate repulsive forces that keep the flywheel magnetically levitating, and thus reduce friction losses within the bearings. The housing of the flywheel should support vacuum conditions that minimize windage losses synonymous with (and proportional to) high rotational speeds. There is also a clutch system that is physically engaged through an electromechanical control system (not shown) to supply/absorb energy from the prime mover whenever the need arises; otherwise, the system keeps rotating at a prescribed constant rotational velocity, ω, that is not expected to decrease or increase without external interaction. The motor/generator connection can either be AC or DC along with the appropriate rectification/inversion circuit.

Another version of flywheels would have the rotor of the motor/generator complete with winding capable of performing the function of the flywheel mass, and thus the

© Springer Nature Switzerland AG 2020
A. H. Alami, *Mechanical Energy Storage for Renewable and Sustainable Energy Resources*,
Advances in Science, Technology & Innovation, https://doi.org/10.1007/978-3-030-33788-9_5

Fig. 5.1 Flywheel examples in **a** potter wheel [1], **b** early steam engine [2], **c** internal combustion engine [3], and **d** pumpjack counterweight in a reciprocating pump

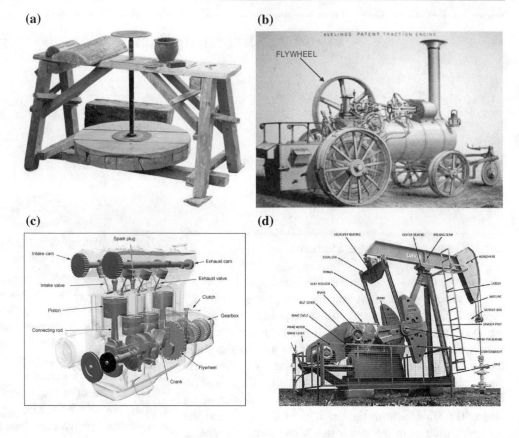

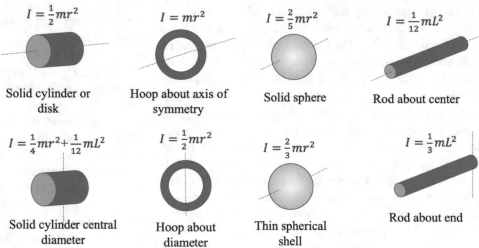

Fig. 5.2 Moment of inertia for common shapes around various axes [4]

$I = \frac{1}{2}mr^2$

Solid cylinder or disk

$I = mr^2$

Hoop about axis of symmetry

$I = \frac{2}{5}mr^2$

Solid sphere

$I = \frac{1}{12}mL^2$

Rod about center

$I = \frac{1}{4}mr^2 + \frac{1}{12}mL^2$

Solid cylinder central diameter

$I = \frac{1}{2}mr^2$

Hoop about diameter

$I = \frac{2}{3}mr^2$

Thin spherical shell

$I = \frac{1}{3}mL^2$

Rod about end

housing would be the stator. This configuration would be more efficient in terms of space saving but would complicate the flywheel construction and maintenance.

The main available technologies for generators/motors (prime movers) that flywheels are expected to play a major role in providing storage for are gas turbines, natural gas engine generators, and fuel cells. In the event of a sudden load (demand) drop in the case of an industrial process turning off, the flywheel steps in and absorbs the load from a fuel cell or other prime mover without disrupting its generation, granting it time to adjust (lower) its output and then the flywheel would disengage either physically (clutch off) or electromechanically if the flywheel was the rotor of the generator. On the other hand, if the load (demand) suddenly increased when an industrial process is suddenly brought online, the flywheel is able to shed some of its power to cover the instantaneous spike in demand, while the prime mover is catching up and would then also disengage. These

Table 5.1 Example of flywheel characteristics [5]

Object	Mass (kg)	Diameter (m)	Angular velocity (rpm)	Energy stored (J)	Energy stored (kWh)
Bicycle wheel	1	0.7	150	15	4×10^{-7}
Flintstone wheel	245	0.5	200	1680	4.7×10^{-4}
Train wheel (60 km/h)	942	1	318	65,000	1.8×10^{-2}
Large truck wheel (60 km/h)	1000	2	79	17,000	4.8×10^{-3}
Train braking flywheel	3000	0.5	8000	33×10^{6}	9.1
Electrical power backup flywheel	600	0.5	30,000	92×10^{6}	26

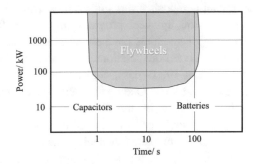

Fig. 5.3 Advantage of flywheels over batteries and capacitors for a specific energy range (power and time) [5]

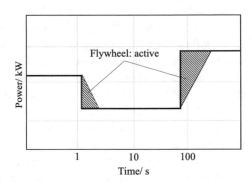

Fig. 5.5 Gradual energy smoothing by flywheel of step load changes [5]

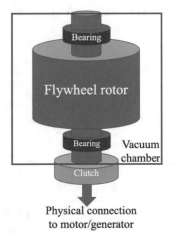

Fig. 5.4 Illustration of the main components of a flywheel storage system

two cases where the flywheel eliminates power disturbances and keeps the voltage constant are shown as the shaded triangles in Fig. 5.5.

5.2 Flywheel Design

In order to achieve maximum kinetic energy absorption/release during the operation of the flywheel, it is more appealing to have faster rotational speeds rather than increase the volume of the flywheel, since the kinetic energy is proportional to the square of ω but (only) directly proportional to I. However, care must be taken during the selection of operational rotational velocities of the flywheel because the linear velocity at the furthest point of the rotating disk equals v = ωr, where r is the radius of the disc, and whenever the velocity approaches the sound velocity within the material, there is the danger of developing dangerous shockwaves that are detrimental to the disk integrity and will cause it to uncontrollably shatter. For example, the speed of sound within steel is ~ 5120 m/s, and if a 1 m flywheel disk is rotating at 10,000 rpm (1047.2 rad/s), the linear velocity of the rim of the disk would be ~ 523 m/s, which is 0.1 M. This should be one of the limiting cases during the material selection phase of the flywheel.

The rapid advances in materials and discovery of reinforced polymers and ceramics have provided flywheel designers with a comprehensive list of materials that can behave favorable under the high values of centrifugal forces that will stem from the high speeds of flywheel rotation. These materials were not available (or at least were not commercially mature) half a century ago and thus the reliance was mostly on metals that were strong enough to handle the load without failure and at the same time contribute to the inertia portion of the energy equation by having high densities or large volumes. Fortunately, this is not the case anymore with the fabrication and implementation of very reliable composite materials such as fiberglass or carbon fiber, that although being very light in weight, are among the strongest materials available, and thus can be spun at high speeds and realize the parabolic augmentation of energy with every step rise of rotational velocity. Table 5.2 lists the mechanical properties of some high-performance fibers that are added to matrix materials to make up composite flywheels. Table 5.3 on the other hand presents the same fibers embedded within an epoxy-based matrix, where the tensile strength of the composite is quoted at the 0° angle of fiber alignment, as well as the fiber strength translation, which is simply the ratio of the strength of the fiber within the composite to the values given for standalone fibers in Table 5.2.

From the tables, carbon fibers with tensile strengths exceeding 6.2 GPa have been developed along with several new polymer fibers such as the new aramids Twaron and Technqa, and the extended-chain polyethylene (PE) fibers Spectra and Dyneema that offer high strengths at a very low density. A rigid-rod polymer fiber, polybenzobisoxazole (PBO), developed at Stanford Research Institute and the Wright-Patterson Laboratory, is available commercially under the trade name Zylon (Tyobo). The consideration of mechanical properties of such fibers on their own would be quite misleading, especially under the high sustained centrifugal loads encountered in flywheel applications. Sometimes, the failure occurs in the matrix materials or at localized defect zones within the material during impregnation or processing, which causes the flywheel to fail at stress levels below the calculated ones, especially under the consistent action of the induced centrifugal force. From Table 5.2, only carbon fiber material offers more than 75% of the original strength of the fiber within the composite, while glass fibers fail to achieve a value higher than 50% regardless of the test protocol. On the other hand, a significant drawback with both carbon and PBO fibers is cost; prices usually range from $30 to over $150 per kilogram (2011 prices).

It should be noted that tensile stress-rupture lifetime behavior of fiber composites was studied most extensively in US labs during the flywheel programs of the 1970s and 1980s. The largest materials database was accumulated by the Lawrence Livermore National Laboratory for the behavior of Kevlar-49, E-glass, S-glass, and AS4 carbon fiber impregnated strands. One of the solutions proposed to enhance the resistance of the flywheel material to the extensive loading would be to introduce compressive radial stresses to the fiber composite flywheels. This is achieved by one of the following techniques: (i) application of programmed tension during filament winding, (ii) using interference fits between concentric cylinders, (iii) bonding of concentric cylinders using an adhesive that is pressurized during cure, (iv) mass loading the inner diameter, and

Table 5.2 Mechanical properties of high-performance fibers [6]

Fiber	Density (g/cm^3)	Modulus of elasticity (GPa)	Tensile strength (GPa)	Manufacturer
T700 carbon	1.80	228	4.83	Toray
T1000G carbon	1.80	297	6.38	Toray
E-glass	2.58	72	3.45	OCF
R-glass	2.55	85	4.33	Vetrotex
S2-glass	2.49	87	4.59	OCF
Fused silica	2.20	69	3.45	J. P. Stevens
Kevlar 49	1.45	120	3.62	DuPont
Kevlar 29	1.44	58	3.62	DuPont
Twaron	1.44	80	3.15	ENKA
Twaron HM	1.45	124	3.15	ENKA
Technora	1.39	70	3.04	Teijin
Spectra 900	0.97	117	2.68	AlliedSignal
Spectra 1000	0.97	173	3.12	AlliedSignal
Dyneema	0.97	87	2.70	Dyneema VOF
Zylon-HM	1.56	269	5.80	Toyobo

Table 5.3 Typical 0° tensile strengths of epoxy-based fiber composites [6]

Fiber	0° tensile strength (GPa)	Fiber strength translation (%)
T700 carbon	2.66	92
T1000G carbon	3.03	70
E-glass	0.8–1.02	39–50
S2-glass	1.63–1.81	59–66
Kevlar 49	1.38–1.52	64–70
Spectra 900	1.09	68
Spectra 1000	1.36	73

(v) building the flywheel using materials that provide a gradient in specific modulus from a low value at the inner diameter to high value at the outer diameter.

Another factor in the flywheel design is its shape. Since a flywheel will be constrained between two bearings, the shape of the flywheel would have to be tapered at the sides and thickening toward the midsection. This design helps in reducing windage losses in case of non-vacuum operation by gradually protruding through air just like a swept back airplane wing. The local thickness of an arbitrary flywheel with maximum radius, r, shown in Fig. 5.6 can be given as

$$b = b_0 \exp[(constant)(r^2)] \tag{5.1}$$

After selecting the material and the shape of the flywheel, the maximum stresses that the design would handle at a certain specific energy limit can be estimated from

$$\frac{E_{kin}}{mass} = \frac{\sigma_{max}}{\rho} K_m \tag{5.2}$$

where σ_{max} is the maximum allowable stress, ρ is the material density, and K_m is the shape factor given in Table 5.4.

Another important consideration in the operation of a flywheel is the applied torque and its effects on the flywheel material. The maximum torque, T_{max}, that can be withstood can be expressed as

$$T_{max} = \frac{2}{3} \pi \sigma_{max} r^3 \tag{5.3}$$

From Newton's second law, the applied torque is related to the change in the rotational velocity by

$$T = I \frac{d\omega}{dt} \tag{5.4}$$

And for a flywheel of radius r and thickness t:

$$T = \frac{\pi}{2} \rho t r^4 \frac{d\omega}{dt} \tag{5.5}$$

And at maximum torque (Eq. (5.3)) this can be rearranged as

$$\frac{d\omega}{dt} = \frac{4}{3} \left(\frac{\sigma_{max}}{\rho r t} \right) \left(\frac{r_{max}}{r} \right)^3 \tag{5.6}$$

Remember that the power is the rate of change of the kinetic energy, so the maximum power is

$$P_{max} = \frac{\pi}{2} \rho \omega t r^4 \frac{d\omega_{max}}{dt} = \frac{\pi}{3} r^3 \sigma_{max} \tag{5.7}$$

Accordingly, the maximum allowable rotational velocity of the flywheel can be estimated to be

$$\omega_{max} = \frac{1}{r} \left(\frac{2\sigma_{max}}{\rho} \right)^{\frac{1}{2}} \tag{5.8}$$

The expected stresses on a flywheel rotor are axial, tangential, and radial. Assuming a cylindrical coordinate

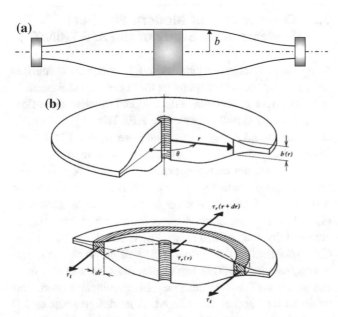

Fig. 5.6 **a** Local thickness, b, variation on a flywheel cross section and **b** the coordinate system used for a flywheel showing the stresses that would develop due to its operation [7]

Table 5.4 Values of the shape factor for several simple disk shapes

Shape	K_m
Brush shape	0.33
Flat disc	0.6
Constant stress disk	1.0
Thin rim	0.5
Thin rim on constant stress disk	0.6–1.0

Table 5.5 Stress summary at various locations of the flywheel [7]

Stress (MPa)/location	Fillet	Center	Interface
@ *rest*			
Radial	−180	−124	−159
Tangential	−138	−124	
Axial	−48	41	
Von-Mises	−165	165	
@ *15,000 rpm*			
Radial	−103	48	−138
Tangential	−69	−48	
Axial	28	28	
Von-Mises	−96	−69	
@ *35,000 rpm*			
Radial	225	303	−41
Tangential	221	303	
Axial	62	−48	
Von-Mises	248	359	

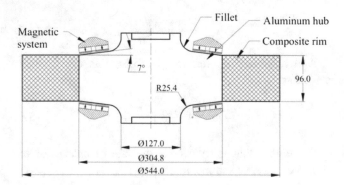

Fig. 5.7 Experimental flywheel stress locations [7]

system, these stresses occur in the *x*-, θ-, and *r*-directions (see Fig. 5.6), respectively. Table 5.5 shows the variation of stresses at various rotational velocities. Von-Mises stresses are representative stress function based on the Von-Mises failure theory and is used extensively in reporting stress (and strain) values. The tested flywheel is shown in Fig. 5.7, indicating the locations of stress measurement.

5.3 Components of Modern Flywheel Systems (Case Study of the AFS TRINITY)

A flywheel storage system, although compact, comprises several independent components that need harmonization in order to arrive at the most effective and efficient operation. For example and in the case of the AFS Trinity [5], the main components and subsystems found are (i) the FMG (Flywheel Motor/Generator) that include a composite rotor, permanent magnet motor–generator, magnetic bearings, and housing, (ii) a control system, and (iii) a power converter. A photo of a cabinet enclosing the system is shown in Fig. 5.8. A brief description taken from the consultant report prepared by the AFS Trinity Power Corporation for the California Energy Commission [5] will be given in the following paragraphs. This report describes each of the main components of the system, as this flywheel system has achieved (and sometimes exceeded) its design goals and is seen to be the future of implementing compact and efficient flywheel power systems for stationary applications.

Fig. 5.8 The M3 AFS trinity flywheel system [5]

5.3.1 Motor/Generator

The AFS TRINITY M3AM incorporates a novel permanent magnet motor–generator invented at Lawrence Livermore National Laboratory and licensed to AFS Trinity. The rotor portion of the motor–generator consists of an array of permanent magnets lining the bore of the rotor. The magnets are oriented to produce a dipole field aligned across the bore of the rotor. The stator consists of turns of finely stranded wire where each strand is insulated (Litz wire). Litz wire is used to minimize eddy current losses in the stator. In discharge operation, the rotating dipole field of the permanent magnets intercepts the windings and induces a voltage in the windings. Conversely, when the machine is charging, the currents impressed on the stator windings create magnetic fields that exert forces on the magnet array in the bore of the rotor, causing it to spin up. The power density and efficiency of the AFS TRINITY motor–generator are exceptionally high, and undesirable thermal loads are substantially avoided.

5.3.2 Bearings

The AFS TRINITY M3AM uses active magnetic bearings (AMB). The M3AM AMB set comprises (1) a stator with radial and axial actuators and radial and axial position sensors; (2) rotor components including laminations, thrust disks, and sensor targets; and (3) a controller system including AMB controller hardware and embedded software.

5.3.3 Composite Rotor

The M3AM rotor material is made of carbon fiber in an epoxy matrix The rotor is about 14 inches long and 10 inches in diameter and operates with a maximum surface speed of about 550 m/s. Prior to use in a flywheel system, a number of rotors are produced and spun-tested to destruction in order to substantiate the safety factor used in the design. Rotors have also been subjected to cycle testing and spin testing at elevated temperature.

5.3.4 Control System

The M3AM system controller is separate and distinct from the magnetic bearing controller. The M3AM control system is comprised of (1) a DSP and associated motherboard, (2) interface components and PCBs for sensors and other communication, (3) fiber-optic communication with the power converter, (4) a user interface with an LCD display and touch pad, and (5) AFS TRINITY designed software driving motor control, voltage regulation, and the user interface. The controller hardware and software represent proprietary AFS TRINITY designs.

5.3.5 Power Electronics

The power electronics include a three-phase H-bridge power converter. The power converter is liquid cooled, and the assembly is surrounded by an enclosure to provide EMI shielding. The power converter communicates with the controller through fiber-optic links to minimize transmitted EMI. The maximum power of the M3AM under normal operating conditions is about 120 kW.

5.3.6 Ancillary Systems

The M3AM incorporates all required subsystems within the flywheel power system enclosure. These include a vacuum system and a liquid cooling system for the stator and power converter. The cooling system uses an electrically nonconducting oil to cool the stator windings directly. Heat from the cooling system is dissipated through heat exchangers located at the top of the M3AM cabinet.

5.3.7 Switchgear

The switchgear system protects both the flywheel system and the application load from various types of fault conditions during operation. A DC bus contactor connects the power converter to end user devices. Relays in series in the system controller trip and open the bus contactor under fault conditions. These contactors disconnect the flywheel from the power converter and connect the flywheel to the dump resistor. The dump resistor is located on top of the cabinet and is sized to safely decelerate the rotor to rest. This configuration assures that the kinetic energy of the flywheel can be dissipated safely even in the event of a total failure of the system controller and power converter.

5.3.8 Flywheel System Testing

For a flywheel, the degree of integration and interdependence between the various subsystems is great. Typically, the performance of a subsystem can only be evaluated when the subsystem is built into the flywheel system. For example, the suitability of a bearing cannot be completely verified unless it is being used to support a flywheel operating at full speed and full power. Similarly, the control system cannot be fully characterized and debugged unless it is tested in conjunction with a fully functioning flywheel.

5.3.9 System Performance

The performance of the M3 Trinity flywheel was based on (i) output voltage, (ii) step load transient response, (iii) start-up and charging time, and (iv) power versus time.

1. *Output Voltage*

In discharge mode, the flywheel system regulates the DC bus voltage to the value set by the user. Output voltage testing demonstrated a voltage control for an output power range from 0 to 100 kW, 50% speed to 100% speed, and bus voltage settings ranging from 580 VDC to 680 VDC. After recovery from an initial transient, the flywheel system

regulated bus voltage to better than ±0.75% under all operating conditions.

2. *Step Load Transient Response*

Immediately after a step change in load, DC bus voltage will momentarily drop below the user-selected value. Transient response measurements were conducted to determine the magnitude of the transient and the duration of the transient for a wide range of operating conditions.

3. *Start-up and Charging Time*

The amount of time required for the system to attain a state of full charge after starting from a dead stop was measured for a range of charge conditions. For representative available charging power, the M3 can reach full charge from a dead stop in less than 4 min.

4. *Power Versus Time*

During discharge, the flywheel system delivers power to the load until extractable energy is depleted. The discharge duration was measured into various loads. Voltage regulation was maintained over the time interval at each power level. Table 5.6 summarizes the output power relation to discharge times, and Fig. 5.9 shows a plot of the results.

5. *Standby Power Consumption*

Standby power is the sum of two terms: power consumed to overcome drag and maintain the flywheel rotor at a particular state of charge, and power that is used by auxiliary systems. The power consumed by the auxiliary system and within the power electronics is distinct from losses that exert drag on the rotor. The total standby power consumption for the present system is 931 W, and that this value will be reduced considerably through straightforward design modifications that are presently planned for the next phase of development of the M3.

6. *Energy Recovery Efficiency*

Energy is lost during the charge–discharge process due to the efficiency of energy conversion of the power converter and the motor. Energy lost in the charge and discharge processes is separate and distinct from standby power consumption and needs to be accounted for separately. Energy

Table 5.6 Voltage-regulated output power versus discharge time [5]

Power (kW)	Discharge time (s)
122.1	10.26
115.6	11.46
108.7	13.06
102.1	14.50
67.6	26.00
33.7	58.50
13.5	150.60

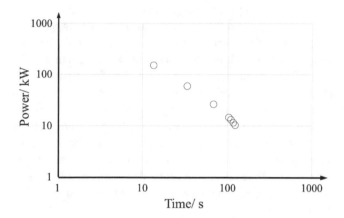

Fig. 5.9 Voltage-regulated output power versus discharge time [5]

recovery efficiency is defined as the fraction of input electrical energy that is retrieved from the flywheel and delivered as output electrical energy. For high power cycle testing performed under this program, energy recovery.

7. *Duty Factor*

A duty cycle is defined as a discharge followed immediately by a recharge. A duty cycle is characterized by the rate of discharge, the depth of discharge, and the rate of recharge. The duty factor characterizes the frequency with which this charge–discharge cycle is repeated. The full cycle passes through three phases: (i) discharge from full speed, (ii) recharge from minimum speed to full speed, and (iii) dwell at full speed.

For high-power energy storage, the duty factor is defined with the following characteristics of the flywheel:

- The full rated power of the flywheel is 100 kW.
- Delivered energy corresponds to a 15-second discharge at rated power (1.5 MJ = 100 kW × 15 s).

A duty factor of 100% is defined as 100 kW, 15-second discharge from full speed, 100 kW, 15-second recharge, and no dwell at full speed. By this definition, operation at a duty factor of 100% corresponds to a repetition of the full power

charge–discharge cycle at intervals of 30 s. Note that the duty factor is defined as a ratio of (30 s)/(actual charge–discharge cycle duration). Then, a duty factor of 50% corresponds to a charge–discharge cycle repeated at intervals of 60 s where the cycle begins with a 1.5 MJ discharge from full speed. The remainder of the cycle comprises an arbitrary combination of recharge and dwell.

For example, a duty factor of 33% implies cycle duration = 90 s, duty factor of 25% implies cycle duration = 120 s, and duty factor of 10% implies cycle duration = 5 min.

Testing conducted under this program demonstrated a sequence of more than 100 consecutive charge–discharge cycles at a duty factor of 40%. Operation at higher duty factor was precluded by facility limitations limiting our maximum charge rate to about 50 kW.

5.3.10 Operational Considerations

With the high centrifugal forces that the flywheels are subject to, close attention has to be paid to the material response to such loads on the output and integrity of the flywheel. The M3 Trinity rotors were built as a single piece by filament-winding glass fiber with successive overwrap layers of medium stiffness carbon fiber and high stiffness carbon fiber. Two significant deficiencies of this composite material and process design were reported by the testing team. First, the relatively large thickness of the part required numerous intermediate cure stages and the length of the curing process was not readily scalable to low production cost. Second, the low elastic modulus of the glass fiber at the bore of the rotor allowed considerable dilation of the bore during operation, which resulted in amplified unbalance, shortened bearing life, and cracking of the magnets. (The NdFeB magnets are mounted in the bore of the rotor and strain follows the material on the bore of the rotor as it grows from centrifugal force during high-speed rotation.) In 1999, the team transitioned into all-carbon construction, where press-fit assembly replaced a gradient in elastic modulus as a way of mitigating matrix material stresses. While the economic benefit of

Table 5.7 Trinity power
flywheel size and power [5]

Parameter	Achieved
Cabinet size, in.	24 × 22 × 78
Cabinet volume, ft^3	23.8
Energy/volume	0.017
Rated power/volume	4.2

shorter curing time was offset by the increased cost of finishing the press-fit rings, all of the technical objectives of this construction have been realized. Bore dilation during operation was reduced, unbalance growth was reduced, and magnet cracking was eliminated. Table 5.7 lists the characteristics of the M3 Trinity flywheel prototype.

5.3.11 M3 Trinity System Technical Specifications

The tested M3 Trinity system is expected to deliver 0.42 kWh at rated power (and more energy at lower power ratings). The resulting ride-through time of 15 s is well matched to the needs of several applications, including load following for distributed generation, ride-through for voltage sags and standby generator start-up, and heavy hybrid electric vehicles. The system power rating is 100 kW, which may not be economically feasible compared to batteries for long discharge times. Although several development groups were engaged in addressing this market in 1998, all the private companies engaged in this endeavor have now come to the same conclusion and ceased development of such a machine. It is at the relatively short discharge times of seconds to a few minutes that most technologists now expect flywheels to play an important role. One of the difficulties in customer understanding of energy storage is that the conclusions on appropriate technology change with the characteristic discharge time of the application: there is no "one size fits all" solution.

As for the system response time to a step change in load, the DC bus voltage regulated by the flywheel system will drop momentarily and then recover to the voltage level that existed prior to the application of the load. The drop will be on the order of 10 s of volts and will have a duration of 10 s of milliseconds with exact values depending on the magnitude of the step load change and the settings entered into the user interface. Response characteristics can be tailored to the application requirement.

Regarding the expected system roundtrip efficiency, it is sometimes viewed from the standpoint of the discharge side of operation. The energy that is stored in the flywheel comes from energy that could otherwise been wasted, and thus the charge leg of efficiency is sometimes not considered in efficiency calculations. This is an interesting perspective, as

some potential flywheel applications require very frequent discharge, others do not. The major source of losses is considered to be the system standby power consumption, and it is what imposes energy costs on the end user.

The ancillary systems used in early prototypes are industrial products with good reliability and would be suitable for use in field test units and commercial products. The cost of this equipment comprises a significant part of the total system cost and development work on ancillary systems will seek to reduce cost by combining functions (rotor system that also performs vacuum pumping) and reduce capacity (less expensive heat exchangers). Ancillary systems are also the only portions of the system that require routine scheduled maintenance, and future development will seek to eliminate required maintenance to the greatest practical extent.

Finally, the M3 flywheel system lifetime would easily reach a cycle life exceeding 10^5 cycles and possibly 10^6 cycles but the developers of the system were unable to execute a program to demonstrate this due to the prohibitively cost of high-cycle spin testing. There is no firm design life limit to the primary flywheel components, and a 20-year life is believed to be a conservative expectation. In practice, power electronics and ancillary systems are expected to have much shorter service lives than the flywheel itself and life improvement will be part of continued engineering of ancillary subsystems.

5.3.12 Potential Market for M3 Flywheels

As for any flywheel developer, the following four application areas are the most important:

1. Power quality/UPS,
2. Distributed generation load following,
3. Industrial pulsed power, and
4. Light rail power management.

1. *Power Quality/UPS*

In this application, the flywheel replaces lead–acid batteries in an uninterruptible power supply (UPS) units. DC power will connect to most commercial UPS power converters in place of the standard battery cabinet. This market is completely addressable market by a flywheel replacing the traditional battery. The market value for this application was around $2.5 billion in 2008.

2. *The distributed generation load following application*

This application enables fuel cells, microturbines, and natural gas engine generators to operate in both grid-connected and isolated (islands) modes with greater reliability and power quality than they could achieve alone. The flywheel technology is uniquely applicable to this situation because it can recharge as quickly as it discharges. The resulting duty cycle capability matches the need for a power management device to act as a power source or sink for step changes in load, and flywheels will facilitate greater customer acceptance of distributed generation. A market value of $3.3 billion (2008) was estimated for distributed generation solutions. Load following is expected to be a small part of the total distributed generation market in the coming years.

3. *The industrial pulsed power application*

This application caters for the needs of a variety of process industries whose machinery creates highly variable or erratic power demands on the grid or their local generation source. Silicon wafer production, arc welding, mining, printing, paper milling, textile milling, and lumber production are examples of such industries. The flywheel is expected to meet the brief power surge requirements for machinery in such industries, thereby avoiding both voltage sags and high utility demand charges. The 2008 market value for this application is $3 billion.

4. *Light rail power management*

This application is similar to the industrial pulsed power and hybrid vehicle applications. Uniquely, it requires many flywheels to be installed trackside at key points in a municipal light rail system. The flywheel provides acceleration power for trains leaving stations or climbing inclines, and it captures regenerative braking energy when trains approach stations or descend grades. A flywheel company, URENCO, has validated this application with flywheel installations at over ten light rail lines around the world, and the Commission will co-fund a demonstration of the URENCO flywheel on the MUNI system in San Francisco. The 2008 market value for light rail power management is $1 billion.

5.4 Flywheels and Regenerative Braking Systems

Another application where flywheel utilization is gaining prominence is in regenerative braking. The simple concept is the momentary storage of the kinetic energy from the engine in revving up the flywheel as the vehicle decelerates instead of depleting it as heat in the brake pads. This energy would allow modern cars to disengage the engine for long waits (idling at a traffic light, for example) and be stored in flywheels mounted on drive wheels. This would benefit in instantly providing the required energy surge to restart the motion of the vehicle while the engine reengages and takes over. The advantage of operating such a system was realized by the Formula One governing body and prompted them to introduce the kinetic energy recovery system (KERS) to Formula One from the 2009 season, as conventional battery (or even hybrid) power systems would have weighed too much and provided too little instantaneous power.

For commercial road vehicles, only Volvo has introduced a solution that can be adopted for commercial vehicles in 2011. The flywheel itself is made of carbon fiber and is housed in a vacuum-sealed casing to keep it free from windage losses. It weighs 6 kg and can spin up to 60,000 rpm and the whole system added only 60 kg to the Volvo S60 model. It consists of the flywheel module, continuously variable transmission (CVT), and a gear set coupling to the rear (driving) axle. Figure 5.10 shows a schematic on the KERS system.

5.4.1 Principal of Operation

During braking, the torque coming from the rear axle spins the flywheel up to speed to store the kinetic energy while the car slows down. When the car needs acceleration, the flywheel is coupled to the CVT and gear set, transferring the kinetic energy back to the rear axle. If the flywheel is fully charged, it can provide a maximum boost of 80 hp for 10 s. The CVT is used to vary the ratio to match wheel speed. When the ratio speeds up the flywheel, it stores energy and conversely, when the ratio slows down the flywheel, it releases energy. A clutch can disengage the flywheel when the car is traveling too fast to avoid overspeeding the flywheel. Derek Crabb, Vice President Powertrain Engineering—Volvo Car Group, is quoted as saying: "The flywheel's stored energy is sufficient to power the car for short periods. However, this has a major impact on

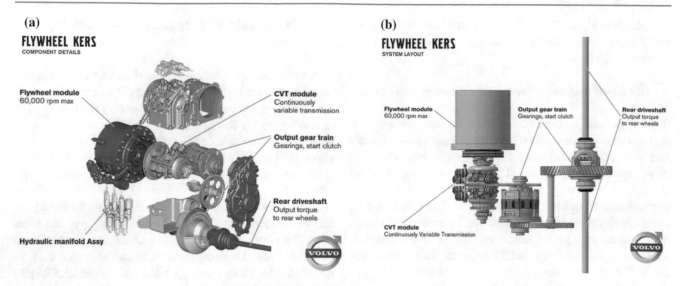

Fig. 5.10 The (**a**) components and system layout (**b**) of Volvo's KERS regenerative braking system using flywheel technology [8]

fuel consumption. Our calculations indicate that the combustion engine will be able to be turned off about half the time when driving according to the official New European Driving Cycle" [9].

If the car does not use the stored energy, the flywheel will slow down gradually due to friction. Since the friction is minimized and there is no air resistance, it takes 30 min to come to a complete stop. However, the kinetic energy drops exponentially; thus, the power lasts much shorter.

5.5 Superconductors and Flywheels

The idea of using the active magnetic bearing system is extremely useful in reducing friction losses as the flywheel is spun up or down. In conjunction with vacuum systems, they both increase the spin-down time of the unutilized energy in a flywheel to tens (if not hundreds) of hours. The concept of superconductivity is of great interest when discussing flywheels, since flywheel systems are electromechanical in nature, and the actual spinning mass is but a single (be it crucial) component of that system. As discussed in the case study of the M3 Trinity flywheel, control components are important for the operation and timely response of any flywheel system. Employing a superconductor is attractive not only in lieu of the active magnetic bearing system but also has the promise of perpetual frictionless rotation that capitalizes on the flux pinning phenomenon to keep the flywheel in place regardless of any physical bumps the flywheel would undergo.

A small introduction will be given here on superconductivity without getting into too much details that could be distracting of the reader, as it would be out of the scope of this book.

Certain materials would experience superconductivity when they are cooled down to a sufficiently low temperature, and their electrical resistance vanishes completely. This is linked to the quantum-mechanical nature of solids, and in particular, to the tendency of electrons to become paired under cryogenic conditions. This is known as "Cooper pairs" that electrons form, where they behave cooperatively in certain materials and form a single quantum-mechanical state. And since physical systems naturally seek a state of lowest energy, the system of particles (electrons in a conductor) will occupy a lowest energy state known as the ground state, unless they are excited by some external source of energy. In certain materials, it is possible for electrons to achieve a ground state with lower energy than otherwise available by entering the superconducting state. Quantum mechanics is called upon to explain this behavior, as it describes physical entities (such as electrons) mathematically by wavefunctions possessing an amplitude (the height of the wave) and a phase (whether it is at a crest or a trough or somewhere in between). These moving waves oscillate both in time and in space, but the catch is that one can either know the location of a particle or know its momentum (where it is heading), but not both. One can sit at one point in space and observe a wavefront that will move up and down in time (undulates in space). This is the unique property of quantum mechanics: the interchangeability of particles and waves in describing physical systems.

For superconductors, there is a unique wavefunction required to describe each particle in a physical system: In "normal" state, wavefunctions describing the electrons in a material are unrelated to one another. In a superconductor, a single wavefunction describes the entire population of superconducting electron pairs. That wavefunction may differ in phase from one place to another within the super-conductor but knowing the function in one place determines it in another. Physicists call such a wavefunction a "many-body wavefunction."

And one should remember that electrons are indistinguishable particles; there is no way to keep track of an individual electron. Pairs of electrons that comprise the superconducting state are constantly forming, breaking, and reforming such that the wavefunction that describes the superconducting state remains the same.

It is also convenient to understand the superconductive behavior of a material in terms of superconducting rings. Consider cooling a ring of superconductor in a small magnetic field that corresponds to one flux quantum threading the ring (a superconducting ring threaded by a single flux quantum). Once the applied field is turned off and according to Faraday's Law of Induction, the moment the field lines that thread the ring changes, a current flows in the ring and this induced current tries to oppose the change in magnetic field by generating a field of its own to replace the one removed. In an ordinary material, that current would rapidly decay away. However, in a superconductor, if the induced current decreased in the ring, the flux threading would be less than a flux quantum and this is not permitted. The next allowable value of flux is zero flux and thus the current would have to abruptly cease rather than decay away. Because the superconducting state is composed of an enormous number of electrons that are paired up and occupying the same quantum state, a current reduction of the sort needed would require all the electrons to jump into another state simultaneously. This is not allowed and as a result, the current induced in a superconducting ring will flow indefinitely. This causes the phenomenon known as flux pinning, and if a magnet interacts with a material in a superconducting state the flux lines will not pass through and the magnet will be pinned in its position as long as the superconducting state endures. This has so many attractive applications for flywheels such as the frictionless spinning of a wheel while levitating on superconductors, or if they are made from a superconducting material themselves.

The challenges remain in using superconductivity for flywheel applications. First, there is the generation of large mechanical forces acting on materials in response to the large magnetic fields present, and the mitigation of this adds considerably to the cost of the system in the form of supports and infrastructure. Second, superconductors have to be maintained below a material-specific critical temperature and the attainment of the required low temperature by the use of a cryo-static refrigerator requires extensive energy and insulation provisions, which is sure to significantly increase the operational costs. Lastly, superconducting materials lose their superconductive property if the value of the surrounding applied magnetic field, H, is above a critical value (critical field). This would cause the superconductive effect to be abruptly lost, and this would give rise to Joule heating (I^2R) once the resistance of the material is restored. Properties of some of the superconducting materials are given in Table 5.8. Note that the higher the critical temperature, the more promise a material has of becoming a high-temperature superconductor [10].

5.6 Flywheels for Energy Storage and Attitude Control of the International Space Station

Satellites are in a state of perpetual fall toward the earth due to microgravity conditions they operate under. In order to prevent an unintended downward spiral toward earth, these satellites employ a number of techniques to remain at a certain orbit away from earth. One technique is the release of propellant (just like an aerosol can), and the other is to use the centrifugal rotation that generates a precession torque about the axis of rotation and in a direction perpendicular to the angular momentum. Flywheels have been used in the International Space Station (ISS) and other satellites that orbit the earth for attitude control via the torques they generate while rotating. With the advances in materials and charge/discharge control modules, the utilization of flywheels to achieve attitude control along with energy storage is an organic extension of using this technology in space.

Table 5.8 Values of the critical temperature and magnetic field for some superconductive materials

Material	Critical temperature (K)	Critical field (T)
Nb-Ti	10	15
Nb_3Al	18	–
Nb_3Ge	23.2	37
Nb_3Sn	18.3	30
NbTi	10	15
MgB_2	39	74
YBCO ($YBa_2Cu_3O_7$)	92	–

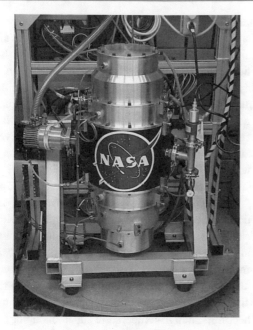

Fig. 5.11 A NASA flywheel with its control system [11]

Batteries are typically used to store and supply electrical energy produced by the extraterrestrial photovoltaic modules attached to the space station; however, heavy batteries occupy large volumes in the spacecraft and their useful life often limits the life of a spacecraft. On the other hand, spinning cylinders, such as reaction wheels and control moment gyroscopes (CMGs), are often employed to control

orientation without expending propellant. Although these devices possess significant rotational kinetic energy, there are no provisions to convert it back into electrical form. Once electromechanical flywheel systems are installed and used for energy storage, they offer an attractive alternative to batteries. Their longevity is superior, and they are less massive than batteries. Moreover, flywheels can simultaneously store energy and perform attitude control tasks, making it possible to reduce spacecraft mass even further.

Each device in the ISS Flywheel Energy Storage System (FESS), formerly the Attitude Control and Energy Storage Experiment (ACESE), consists of two counterrotating rotors placed in vacuum housings and levitated with magnetic bearings. The compact setup is shown in Fig. 5.11. The subcomponents are also shown in Fig. 5.12.

Motor–generators will connect the rotors to the existing electrical power system so that they can store energy when it is available from the photovoltaic arrays, and supply energy when it is needed. Each rotor, made up of a metallic hub and a rim of composite material, is approximately 28 cm in diameter and 38 cm in length, and spins at angular speeds ranging between 18,000 and 60,000 rpm. Whereas the attitude control function of the ACESE was to have been performed with one pair of rotors, as many as 48 pairs will make up the FESS on the "assembly complete" configuration of ISS.

The characteristics of such flywheels exhibit their promise. With a specific energy (specific energy is at the system level, and a system is defined to include the flywheel

Fig. 5.12 Subcomponents of NASA flywheel [12]

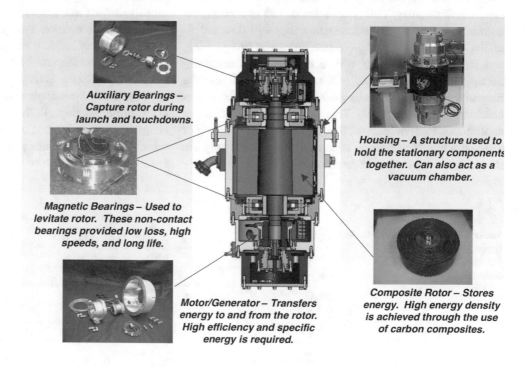

Auxiliary Bearings –
Capture rotor during
launch and touchdowns.

Magnetic Bearings – Used to
levitate rotor. These non-contact
bearings provided low loss, high
speeds, and long life.

Motor/Generator – Transfers
energy to and from the rotor.
High efficiency and specific
energy is required.

Housing – A structure used to
hold the stationary components
together. Can also act as a
vacuum chamber.

Composite Rotor – Stores
energy. High energy density
is achieved through the use
of carbon composites.

Table 5.9 NASA rotor development for flywheel energy storage onboard the ISS [12]

Flywheel	HSS	Dev1	D1	G2	FESS	G3
Features	Steel hub	Single-layer composite	Multilayer composite 750 m/s	Multilayer composite 750 m/s	Multilayer composite 950 m/s	Composite arbor 1100 m/s
Energy (Wh)	17	300	350	581	3000	2136
Specific energy/ (Whkg^{-1})	1	23	20	26	40	80
Life (yr)	NA	<1	1	1	15	15
Temperature (°C)	NA	NA	25–75	25–75	NA	−25–95
Illustration						

modules, power electronics, sensors, and controllers) of 25 Wh/kg, and an efficiency of 85% (efficiency is also measured at the system level as the ratio of energy recovered in discharge to energy provided during charge), a lifetime of around 15 years (compared to a couple of years for batteries), and a tested operational temperature range of −45° to 45 °C.

The progress in flywheel development for NASA is quite interesting. Table 5.9 shows the development of various rotor materials that serves the purpose of providing attitude control of the ISS as well as energy storage. Starting from the classical steel hub flywheels (that is added for reference), the composite material flywheels made it far more attractive to install flywheels instead of the more conventional battery banks. This is a major benefit with the intermittent availability of the solar photovoltaic modules. Flywheels have flexible charge/discharge profiles, so these solar arrays are more fully utilized. More importantly, flywheels can operate over extended temperature ranges, reducing thermal control requirements that would arise with chemical battery banks, and their state of charge is precisely known from their velocity profiles, thus limiting any unpleasant surprises and power outages, especially toward the end-of-life period of batteries.

References

1. How pottery was made (2013), www.howitworksdaily.com/how-pottery-was-made/
2. V. Ryan, FLYWHEELS (2017), http://www.technologystudent.com/energy1/flywheel1.html
3. The basics of 4-stroke internal combustion engines, in *Xorl* (2011), https://xorl.wordpress.com/2011/03/05/the-basics-of-4-stroke-internal-combustion-engines/
4. R. Nave, Rotational-linear parallels, http://hyperphysics.phy-astr.gsu.edu/hbase/mi.html
5. G. Newsom, Flywheel systems for utility scale energy storage (2019), https://ww2.energy.ca.gov/2019publications/CEC-500-2019-012/CEC-500-2019-012.pdf. Accessed 22 July 2019
6. S.J. Deteresa, S.E. Groves, Properties of fiber composites for advanced flywheel energy storage devices
7. A. Filatov, P. Mcmullen, K. Davey, R. Thompson, Flywheel energy storage system with homopolar electrodynamic magnetic bearing*
8. Hybrid cars (2017), https://www.autozine.org/technical_school/engine/Hybrid.html
9. Volvo car corporation developing flywheel kinetic energy recovery system; considering broad application - green car congress, in *Green Car Congress* (2011), https://www.greencarcongress.com/2011/05/vcc-20110526.html. Accessed 12 Sept 2019
10. A.H. Alami, C. Aokal, M.A. Assad, Facile and cost-effective synthesis and deposition of YBCO superconductor on copper substrates by high-energy ball milling. Metall. Mater. Trans. A **12**, 1073–5623 (2016)
11. C.M. Roithmayr, International space station attitude motion associated with flywheel energy storage, https://ntrs.nasa.gov/search.jsp?R=20040086750. Accessed 12 Sept 2019
12. Development of a high specific energy flywheel module, and studies to quantify its mission applications and benefits Tim Dever/NASA GRC, https://ntrs.nasa.gov/search.jsp?R=20150009522. Accessed 15 Aug 2019

6.1 Pumped Hydro Storage

Pumped hydro storage is analogous to the operation of a massive battery, capable of storing hundreds of megawatts of energy in a simple and sustainable manner. Hydrogeneration projects are strategic in nature and always involve an investment on a national scale. Hydroelectric power is the gift of nature for countries that have water resources (rivers or lakes) and also the topography that supports such projects.

Pumped hydro storage is complementary to hydroelectric generation, and its concept of operation is quite simple, as shown in Fig. 6.1. During periods of high demand, water from the upper reservoir is released with large momentum through water turbines, where the substantial water head stored behind dam walls converts the potential energy into mechanical energy through the rotation of the turbines. The turbines are coupled with generators that produce steady electrical supply that can quite comfortably cover the baseload requirement without interruption. This operation phase, shown in Fig. 6.1, pertains to the hydrogeneration part of such an installation.

The pumped hydro storage part, shown in Fig. 6.2, initiates when the demand falls short, and the part of the generated electricity is used to pump water from the lower reservoir back into the upper reservoir. Since this operation is allowed to take place for a time duration from six to eight hours (before the demand surges up again the next day), the power used up by the pump motors is not expected to be substantial and is only a fraction of the power generated via potential energy conversion. Pumps only need to have enough energy to surmount the elevation head demand, and the rate of energy consumed can extend overnight. One can think of this in terms of energy needed to climb a flight of stairs. You can either go up one step at a time, or you can run up the stairs, skipping two steps at a time. In both cases, the energy required is the same, but the power expended (energy/time) is different. And thus, such installations capitalize on the relatively long periods of no demand to secure the water back into the upper reservoir.

It is also important to note that run-of-river systems share the same operational concept as shown in Fig. 6.1, with no need of a reservoir to hold the water at a certain elevation. These systems capitalize on the kinetic energy already provided by the flow of the river (or more preferably a waterfall). For example, Niagara Falls Hydropower Plant on both the Canadian and the American sides have a height of 51 and 54 m and 600,000 and 2,400,000 L per second, respectively. Figure 6.3 shows the location of the Niagara Falls power station, along with a cross section of the enormous original 70,000 hp cast iron water turbine coupled with the electric generators. As the river and waterfall are moving continuously, pumped hydro storage options seem infeasible, as there is little benefit from pumping the water back into the upper reservoir, and perhaps other storage technologies (like flywheels) can be employed.

Table 6.1 lists some examples of large pumped hydro stations around the world. It is interesting to note that the Niagara Falls station (and other run-of-river stations) is not mentioned there as they are not considered pumped hydro storage facilities.

6.2 Governing Equations

Fluid flow is governed by the complex Navier–Stokes equations that are partial differential equations that originate from Newton's second law, $\vec{F} = m\vec{a}$, and include all types of forces that affect a fluid element, with all resultant accelerations that are induced. In contrast to solids, the Second Law applied to fluids involves the inclusion of a host of forces, such as body forces, friction (viscous) forces, and pressure forces that have multiple effects on the body resulting in more than mere translation of the particle. Remembering that a particle velocity is the undisputed indicator of its direction of motion, and that accelerations resulting from applied forces pertain not only to changes in the magnitude of the

© Springer Nature Switzerland AG 2020
A. H. Alami, *Mechanical Energy Storage for Renewable and Sustainable Energy Resources*,
Advances in Science, Technology & Innovation, https://doi.org/10.1007/978-3-030-33788-9_6

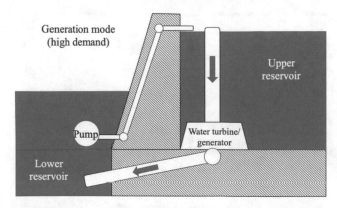

Generation mode (high demand)

Upper reservoir

Pump

Water turbine/ generator

Lower reservoir

Fig. 6.1 Generation operation mode where water is routed through the turbines

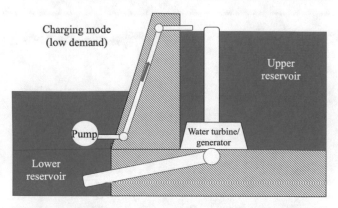

Charging mode (low demand)

Upper reservoir

Pump

Water turbine/ generator

Lower reservoir

Fig. 6.2 Charging operation mode where water driven by pumps from lower reservoir to upper reservoir

(a)   **(b)**

Fig. 6.3   **a** Niagara falls run-of-river hydropower station with **b** the 70,000 hp ~50 m head 107 rpm water turbine [1]

velocity but also changes in the direction of motion, too, then one can expect an extremely complicated vector equation to govern fluid flow. One should rely on a solid mathematical model in order to predict the effect of the forces applied on the fluid as a whole (and not just the constituent particles). Remember that in fluid mechanics, Navier–Stokes is also known as the energy equation, and in order to isolate the energy exchange with a fluid, one should describe the system in terms of these equations.

Just to appreciate the magnitude of the computational power that has to be availed to solve these equations (no one has successfully managed to solve or apply these equations analytically), they are reproduced in the following simplified vector form shown in Fig. 6.4.

The equation in Fig. 6.4 can be resolved in the *xyz* cartesian space and used in conjunction with the continuity equation, momentum equation, and the energy equation as follows:

The continuity equation:

$$\frac{\partial \rho}{\partial t} + \frac{\partial(\rho u)}{\partial x} + \frac{\partial(\rho v)}{\partial y} + \frac{\partial(\rho w)}{\partial z} = 0 \qquad (6.1)$$

The equations for x-momentum:

$$\frac{\partial(\rho u)}{\partial t} + \frac{\partial(\rho u^2)}{\partial x} + \frac{\partial(\rho uv)}{\partial y} + \frac{\partial(\rho uw)}{\partial z}$$
$$= -\frac{\partial p}{\partial x} + \frac{1}{Re}\left[\frac{\partial \tau_{xx}}{\partial x} + \frac{\partial \tau_{xy}}{\partial y} + \frac{\partial \tau_{xz}}{\partial z}\right]$$
$$(6.2)$$

The equations for y-momentum:

$$\frac{\partial(\rho v)}{\partial t} + \frac{\partial(\rho uv)}{\partial x} + \frac{\partial(\rho v^2)}{\partial y} + \frac{\partial(\rho vw)}{\partial z}$$
$$= -\frac{\partial p}{\partial y} + \frac{1}{Re}\left[\frac{\partial \tau_{xy}}{\partial x} + \frac{\partial \tau_{yy}}{\partial y} + \frac{\partial \tau_{yz}}{\partial z}\right]$$
$$(6.3)$$

Table 6.1 Examples of pumped hydropower stations around the world [2]

Country	Station name	Capacity (MW)
Argentina	Rio Grande-Cerro Pelado	750
Australia	Tumut Three	1,500
Austria	Malta-Haupsufe	730
Bulgaria	PAVEC Chaira	864
China	Guangzhou	2,400
France	Montezic	920
Germany	Goldisthal	1,060
	Markersbach	1,050
India	Purulia	900
Iran	Siah Bisheh	1,140
Italy	Chiotas	1,184
Japan	Kannagawa	2,700
Russia	Zagorsk	1,320
Switzerland	Lac des Dix	2,099
Taiwan	Mingtan	1,620
United Kingdom	Dinorwig, Wales	1,728
United States	Castaic Dam	1,566
	Pyramid Lake	1,495
	Mount Elbert	1,212
	Northfield Mountain	1,080
	Ludington	1,872
	Mt. Hope	2,000
	Blenheim-Gilboa	1,200
	Raccoon Mountain	1,530
	Bath County	2,710

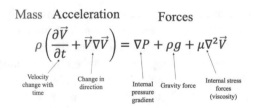

$$\rho\left(\frac{\partial \vec{V}}{\partial t} + \vec{V}\nabla\vec{V}\right) = \nabla P + \rho g + \mu\nabla^2\vec{V}$$

Mass   Acceleration   Forces

Velocity change with time   Change in direction   Internal pressure gradient   Gravity force   Internal stress forces (viscosity)

Fig. 6.4 Components of the Navier–Stokes equation as vectors

The equations for z-momentum:

$$\frac{\partial(\rho w)}{\partial t} + \frac{\partial(\rho uw)}{\partial x} + \frac{\partial(\rho vw)}{\partial y} + \frac{\partial(\rho w^2)}{\partial z}$$
$$= -\frac{\partial p}{\partial z} + \frac{1}{Re}\left[\frac{\partial \tau_{xz}}{\partial x} + \frac{\partial \tau_{yz}}{\partial y} + \frac{\partial \tau_{zz}}{\partial z}\right] \quad (6.4)$$

The energy equation:

$$\frac{\partial(E)}{\partial t} + \frac{\partial(uE)}{\partial x} + \frac{\partial(vE)}{\partial y} + \frac{\partial(wE)}{\partial z}$$
$$= -\frac{\partial(up)}{\partial x} - \frac{\partial(vp)}{\partial y} - \frac{\partial(wp)}{\partial z} + \frac{1}{RePr}\left[\frac{\partial q_x}{\partial x} + \frac{\partial q_y}{\partial y} + \frac{\partial q_z}{\partial z}\right]$$
$$+ \frac{1}{Re}\left[\frac{\partial}{\partial x}\left(u\tau_{xx} + v\tau_{xy} + w\tau_{xz}\right) + \frac{\partial}{\partial y}\left(u\tau_{xy} + v\tau_{yy} + w\tau_{yz}\right)\right.$$
$$\left.+ \frac{\partial}{\partial z}\left(u\tau_{xz} + v\tau_{yz} + w\tau_{zz}\right)\right] \quad (6.5)$$

Note that the vector equations are resolved in x, y, and z spatial coordinates and the time, t. There are six dependent variables, the pressure p, density ρ, and temperature T, and three components of the velocity vector; the u component is in the x-direction, the v component is in the y-direction, and the w component is in the z-direction. All of the dependent variables are functions of all four independent variables. The differential equations are therefore partial differential equations and not ordinary differential ones. Re is the Reynolds number which is a parameter that is the ratio of the scaling of the inertia of the flow to the viscous forces in the flow. The q variables are the heat flux components, and Pr is the Prandtl number which is a parameter that is the ratio of the viscous stresses to the thermal stresses. The τ variables are components of the stress tensor, where the dynamic viscosity, μ, of the fluid shows up as $\tau = \mu \frac{\partial u}{\partial y}$ in the x-direction. A tensor is generated when you multiply two vectors in a certain way. The velocity vector has three components; the stress tensor has nine components. Each component of the stress tensor is itself a second derivative of velocity components.

Then came Daniel Bernoulli, who assumed that an ideal fluid would have no friction losses (inviscid flow with zero viscosity), and approximated Navier–Stokes equations for incompressible (constant density) and one-dimensional (any variation in y or z is neglected) flow. This enabled the simplification of the fluid flow governing equations to be written as energy heads (static, dynamic, and elevation), which not only has been used for decades, but is behind the successful design and implementation of numerous hydraulic projects and installations, from large-scale piping systems to transport water, oil, and even gas to fire-fighting systems and sewage lines. The following equation is a marvel of simplicity, compactness, and effectiveness:

$$\frac{p}{\rho g} + \frac{v^2}{2g} + z = constant \qquad (6.6)$$

It can quickly be recognized that the terms in this equation can be applied for the pumped hydro storage case, where the elevation can be correlated with the values of fluid velocity that is expected to turn the water turbines in order to generate electricity. The head equation ($z = \Delta H$) can be easily transformed into power units by knowing the volumetric flow rate of the fluid, Q [m³/s], and its specific weight, $\gamma = \rho g$ [N/m³]:

$$P = \gamma Q \Delta H \qquad (6.7)$$

And thus, the pumping power (with a pump efficiency, η) required to raise the water back into the upper reservoir can be estimated from

$$P = \frac{\gamma Q \Delta H}{\eta} \qquad (6.8)$$

And the power generated in a turbine with efficiency η is calculated as

$$P = \eta \gamma Q \Delta H \qquad (6.9)$$

The product of the total volume of water and the head difference between the reservoirs is proportional to the energy stored. For example, with an elevation change of 340 m and a reservoir volume of 12 million m³, approximately 9,000 MWh can be provided to the power grid assuming a generating efficiency of 90%.

Before moving any further with the analysis of the hydroelectric system, it should be noted that Bernoulli's Eq. (6.6) is not self-sufficient because it is too ideal. Assuming that the fluid suffers no losses due to friction does not mean that it does not in real-life applications. And in order not to sacrifice the simplicity and convenience offered by the use of Bernoulli's equation, it would be useful to estimate the losses a fluid incurs, and then either settle for them or add a pumping device to compensate for them. Bernoulli's equation would then look like

$$\frac{p}{\rho g} + \frac{v^2}{2g} + z = constant + \text{friction loss head}, h_f \qquad (6.10)$$

The friction head loss is estimated by the Darcy–Weisbach equation:

$$h_f = \left(f \frac{L}{D} \right) \left(\frac{v^2}{2g} \right) \qquad (6.11)$$

where h_f = head loss (m), f = friction factor, L = length of pipework (m), d = inner diameter of pipework (m), and v = velocity of fluid (m/s).

This is the reason why you were introduced to the Moody chart that have graphically presented work of Petukhov (1970) in his explicit first equation to determine the friction factor in smooth tubes at a certain Reynolds number:

$$f = (0.790 \ln Re - 1.64)^2 \qquad (6.12)$$

This equation is for smooth pipes (plastic pipes, for example) and works for a Reynolds Number range $10^4 < Re < 10^6$. Of course, not all pipes are made of plastic, and thus if there is a roughness, ϵ, in the pipe of diameter D, then it has to be accounted for via Colebrook equation, which estimates the friction factor for turbulent flow to be

$$\frac{1}{\sqrt{f}} = -2.0 \log \left(\frac{\epsilon/D}{3.7} + \frac{2.51}{Re\sqrt{f}} \right) \qquad (6.13)$$

This is not an explicit equation of f, and thus it would be very difficult to solve for f directly. The Moody charts are curves that relate the friction factor to a certain value of Re and also surface roughness; unless the flow was laminar (Re < 2300), then the friction factor is a sole function of Re ($f = 64/Re$). Figure 6.5 depicts the Moody diagram, which should be accessed with the knowledge of Re and the roughness of the pipe. Representative values of the internal roughness of various pipe materials are given in Table 6.2. It should be noted that uncertainty in these values is around ±60%, and a profilometer should be utilized to accurately measure a specific pipe roughness.

Fig. 6.5 The moody chart for friction factor calculations for flow pipes

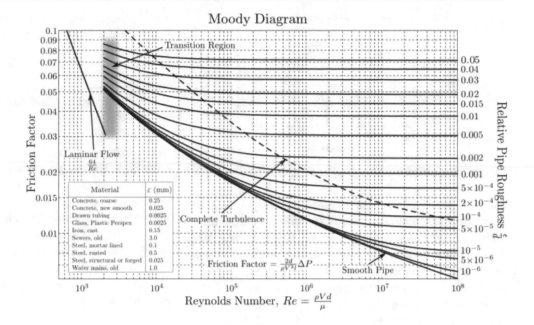

Table 6.2 Representative values of pipes roughness for various materials

Material	Roughness, ϵ/mm
Glass, plastic	0 (Smooth)
Concrete	0.9–9
Wood stave	0.5
Rubber	0.01
Copper/brass	0.0015
Cast iron	0.26
Galvanized iron	0.15
Wrought iron	0.046
Stainless steel	0.002
Steel	0.045

The reason why the previous discussion on fluid mechanics is necessary is because although the analysis of pumped hydro storage systems is straightforward, many basic practical aspects about fluid flow should be well understood in order to properly design and/or operate such facilities at the highest performance levels and select the size and power of turbines and pump to minimize energy losses whenever possible.

6.3 Technical Consideration Pumped Hydro Storage Systems

For a particular site to be favorable for pumped storage hydropower, there are some key technical considerations to be assessed. They include the following:

1. Topographic conditions that provide sufficient water head between upper and lower reservoirs,
2. Strong geotechnical site where no avalanches or landslides expected,
3. Availability of sufficient quantities of water, and
4. Access to electrical transmission networks.

It is also important to design a hydroelectric system with provisions for pumped hydro storage when possible. This should also be done with the consideration of long-term variation of power demand/supply patterns in mind to facilitate the expansion of the operation of a hydro facility in the future. The minimum practical head for an off-stream pumped storage project is generally around 100 m, with higher heads being preferred. Some projects have been built with heads exceeding 1000 m. These projects involve the

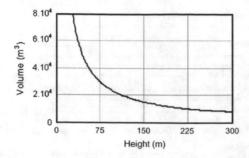

Fig. 6.6 Water volume needed at a given height to store 6 MWh [2]

use of separate pumps and turbines for the pumping and generation operations, respectively, or the utilization of multiple-stage pump/turbines to minimize operational losses resulting from equipment overload. The volume of water available is also an important factor, and hence the need of a permanent and dense water supply to back up the system operation. Figure 6.6 shows the water flow needed versus water head height required to store 6 MWh of energy.

Flow rate of water is another important design factor, and it is determined to achieve a desired cycling time of generation/pumping. Higher flow rates lower the cycling time and require larger size of the generating and pumping units, and waterways diameter. Benefit–cost optimization is generally carried out to optimize the design flow rate, and hence the plant capacity. The optimum flow rate is constrained by the head loss associated with particular waterways diameter and also the cost of the system. The economic diameter of waterways is optimized by balancing the loss of energy benefits due to higher head losses associated with smaller diameters versus waterways construction costs associated with larger diameters.

The reservoir size is another important design parameter, but it is mostly dependent on the geological location. Both upper and lower reservoirs have usually enough storage to

provide generation at full capacity for about six to eight hours, as this is the amount of time where low demand is expected.

6.4 Pumped Hydro Storage System Efficiency

The efficiency of pumped hydro storage facility is usually quite high. The overall efficiency is a function of each of the efficiencies of the component in the system. Data for past decades of operating large stations in the United States show the reported efficiencies to be between 60 and 80% for years 1963–1995. With access to more efficient components (pumps, turbines, pipe network design), it is now possible to achieve efficiencies as high as 80%. Table 6.3 shows representative values of component efficiency for a typical pumped hydro storage plant.

6.4.1 Major Losses for Pumped Hydro Storage Systems

Mechanical and electrical components are not the only culprits contributing to the decline of the system efficiency. The following are other major contributors to overall system losses:

Reservoir evaporation

Evaporative losses depend on the size and location of reservoirs. Shallow reservoirs located in tropical climates and with a large surface to storage ratio are far more impacted by evaporative losses than reservoirs in moderate climates. Similarly, a large shallow reservoir will evaporate

Table 6.3 Representative cyclic efficiency values of a pumped hydro storage plant [3]

	Component	Indicative value (%)
Pump cycle	Water conductors	98.0–98.6
	Pump	90.0–92.0
	Motor	97.8–98.3
	Transformer	99.0–99.6
	Overall	*85.4–88.8*
Generate cycle	Water conductors	98.6–98.0
	Turbine	75.0–91.0
	Generator	97.8–98.3
	Transformer	99.0–99.6
	Overall	*71.6–86.4*
Overall	Losses and Leakage	98.0–99.8

faster than a small and deep reservoir. Evaporation is extreme in conditions of dry heat and wind. Whenever evaporative losses are significant, supplemental water supply may be required to refill some of the reservoir volume. Some innovative solutions are proposed to mitigate this problem, including the deployment of shade balls that float on top of the reservoir to limit the radiation reaching the surface and cause evaporation, as shown in Fig. 6.7.

Leakage losses

Depending on geological conditions, a liner may be required in one or both reservoirs to prevent leakage. Seepage through the reservoir liner could still occur, although lining systems may include a leak detection provision coupled with a seepage collection system designed to capture water lost through the lining, if it occurs. A main source of leakage is cracks that develop in concrete-lined sections of the waterways.

Transmission Losses

Electric power transmission losses are a function of transmission line length, voltage, and conductor size and type. Planning is important to take into account several transmission interconnection options. The selection of a point of connection could involve a study of whether the point of connection should be a nearby substation or if it should be connected to an existing transmission line.

6.4.2 Response Time for Pumped Storage

As expected, the operation generation mode is similar to that of a conventional hydro generator operation. The output of a hydro generator can be adjusted by changing wicket gate opening. Changing gate opening changes the amount of water passing through the turbine, and this capability allows turbine units to be used for automatic generation control and to regulate frequency and load when the plant is in generation mode. Operating a single-speed pump-turbine unit in a regulating mode as a generator results in considerable losses of efficiency. In the pump mode, the unit operates at the gate openings that allow the most efficient operation for a given head.

Some values for typical turnaround and starting times for reversible pump-turbine units are as follows:

- From pumping to full-load generation: 2 to 20 min,
- From generation to pumping: 5 to 40 min,
- From shutdown to full-load generation: 1 to 5 min, and
- From shutdown to pumping: 3 to 30 min

With adjustable-speed machines, it is possible to reduce some of these times since synchronization can occur at a lower speed. The control systems can match the rotor electrical speed and the system frequency within seconds. Synchronization can thus occur more rapidly and well before a machine reaches full speed. Further, when the adjustable-speed machine is in the pumping mode, speed

Fig. 6.7 Shade balls in Ivanhoe Reservoir in Silver Lake [3]

does not need to be brought to or even near its nominal speed for synchronization. A potential reduction in time of 5 to 15% can usually be achieved.

6.5 Operational Aspects of Pumped Hydro Storage Systems

The traditional mode of operation for a pumped hydro storage plant is to pump sometime after 10 PM through midnight and into the early morning hours, during the period when low-cost pumping energy is available from baseload units, and to generate during daytime peak periods when energy values are highest. Since the electrical demand is usually less during weekends, there is more low-cost pumping energy available. In a weekly cycle, the upper reservoir is full on Monday morning (beginning of the week) and nearly empty on Friday evening and the units generate during the week as part of the weekly dispatch. During the weekend, the units pump to completely refill the upper reservoir. The pumped storage operation shown in Fig. 6.8 is used to supply the power for peaks of the system load.

It is clear from the figure how the operation of a pumped hydro facility is analogous to that of a chemical battery. The massive difference is the massive scale difference of energy/power stored.

By consulting the Ragone plot that depicts the pumped hydro system compared with other energy storage systems, its advantage can be clearly seen. One should recall from earlier discussion that the area that is sought in such diagrams is the top-right corner, and from Fig. 6.9 it is seen that it is where pumped hydro systems reside.

The location of the pumped hydro systems at the top-right corner derives from the system ability to operate at different discharge rates depending on demand. The main advantage of

adjustable-speed units is their ability to vary the rotational speed of the pump and turbine prime movers for more efficient operation and better integration with the power grid. The operation of adjustable-speed units varies based on the size of the unit. For units less than 50 MW, a conventional synchronous generator is linked to the power grid by a static frequency converter, while larger units, over 50 MW, are operated in adjustable mode via the application of a doubly fed induction machine (DFIM), with a three-phase sinusoidal rotor voltage and current that is provided by an with a three-phase sinusoidal rotor voltage and current that is provided by an AC/DC/AC solid-state converter. Figure 6.10 shows the difference in electrical connections and rotor type for the single-speed system and variable (adjustable) speed system.

The main components of the motor/generator assembly for the adjustable-speed unit (Fig. 6.10b) are a three-phase stator, three-phase cylindrical rotor, and three-phase collector assembly. The rotor consists of a laminated steel cylinder with slots for three-phase windings that are connected to an external exciter by slip rings mounted to the top of the shaft. The stator winding is energized by the constant frequency of the bulk power system to establish a synchronous alternating magnetic field. Another main difference is that the rotor of the single-speed units (Fig. 6.10b) is excited by a DC source while the rotor of the adjustable-speed unit is excited from a variable frequency voltage source. When the frequency and phase angle of the rotor excitation is adjusted above or below direct current, the rotor speed can be synchronized accordingly. The operation of the adjustable-speed system is a big improvement over single-speed units. The increased operational range allows adjustment of input power which helps to avoid reverse flow when operating at large input heads, controls electrical power frequency of the power grid during pumping mode, and helps to avoid cavitation when operating at low heads.

Fig. 6.8 Representative operating cycle of a pumped hydro storage plant [4]

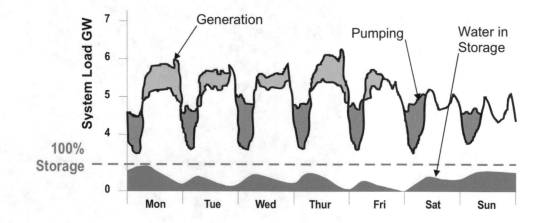

Fig. 6.9 Ragone plot depicting the location of pumped hydro storage systems [5]

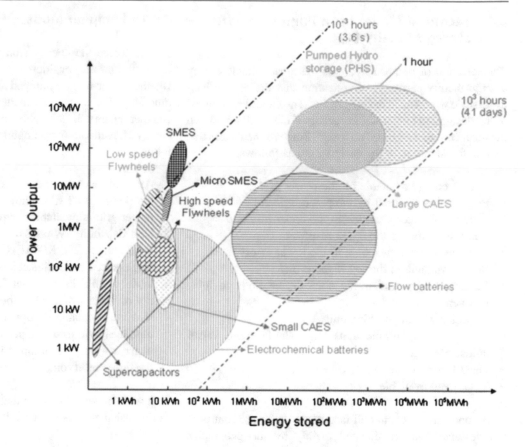

Fig. 6.10 Comparison between **a** single-speed and **b** adjustable-speed electrical drives

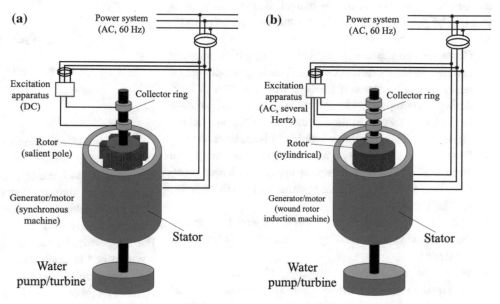

6.6 Technical Model for a Pumped Hydro Storage Facility [4]

The selection of pumped hydro as the most suitable energy storage facility carries the implication that the energy handling scale would be in the megawatts. The variables involved in the design are numerous, especially if one will not rely on the ideal case that assumes no losses. Some of the factors that require the attention of the designer are as follows:

- Excess energy available on the power grid,
- Peak energy required by the power grid,
- Volume of water available in the upper reservoir,
- Space available in the upper reservoir,
- Volume of water available in the lower reservoir,
- Space available in the lower reservoir,
- Efficiency of pump/turbine units during both pumping and generating modes,
- Capacity of pump/turbine units,
- Type of pump/turbine units (e.g., single speed versus adjustable speed),
- Head loss in waterway, and
- Total head available.

As one can notice, not all factors are under the control of the designer and are dictated mainly by the geographic location and also the available budget. Some assumptions that can be made to simplify the calculations are as follows:

- Constant head assumes the total head during both pumping and generating modes is constant. This is generally only a reasonable assumption if there is little variation in head during the pumping cycle. If there is a significant variation in head with respect to time, during pumping and turbine mode, then this assumption would not be reasonable.
- Evaporation and seepage are negligible, assuming that there will be minimal losses in upper and lower reservoirs due to reservoir lining systems and small reservoir surface areas.
- The waterway conduits are adequately sized to prevent the waterway from creating a discharge bottleneck or constraint.
- Waterway losses during pumping and generating mode are equal to 3% of the total head. This includes all pipe friction losses and minor system losses. To achieve better turbine efficiencies, the pump/turbine units are assumed to operate between 70 and 100 percent of their rated capacity during generation mode.
- Typical "off-stream" pumped storage hydropower configuration. A pumped-back configuration would require additional inputs such as stream inflows, required minimum environmental (stream) flows, etc.

6.6.1 Pumping Mode

This is where the energy storage takes place. In pumping mode, the model considers a number of factors to establish whether water will be pumped to the upper reservoir at each time step. There are four main limiting factors that dictate whether energy is available, and if so, what quantity of energy is available for pumping. The limiting factors are as follows:

1. Available Excess Energy: Is excess energy available on the power grid? If so, how much excess energy on the power grid is available for pumping?
2. Unit Capability: What is the capacity of the pump/turbine units (e.g., 200 MW)? Can the designer consider single-speed and adjustable-speed pumps separately?
3. Water Available in Lower Reservoir: What is the quantity of water available in the lower reservoir?
4. Volume Available in Upper Reservoir: What is the volume available in the upper reservoir? At each time step water can only be pumped if there is space available in the upper reservoir.

It was shown previously that the available excess energy can be calculated using the following power equations:

$$E_{pump} = P_{pump} \cdot t \qquad (6.14)$$

$$P_{pump} = \frac{\gamma Q \Delta H}{\eta} \qquad (6.15)$$

Solving for the flow rate, Q,

$$Q = \frac{P_{pump}\eta}{\gamma \Delta H} \qquad (6.16)$$

For example, consider a situation where power available for pumping is 1,000 MW, pumping efficiency is 98%, and net head is 443 meters. Using Eq. (6.16), the available excess energy limiting factor is 226 m^3/s.

Unit capabilities can be calculated for either a single-speed pump or variable speed pump (See Fig. 6.10). For single-speed units, the power input required is almost constant and is easily calculated from Eq. (6.15). For variable speed units, the pumps are capable of operating at a broader range of power inputs and thus could result in more than one value of the required flow rate of water to be pumped (can still use Eq. (6.15) for the full range of P_{pump}).

The next constraint is the water available in lower reservoir. The volume of water available, V, in the lower reservoir is converted into units of m^3/s using the following equation where t is time:

$$V = Q.t \qquad (6.17)$$

which is converted into time available for pumping by

$$Q = \frac{V}{t} \qquad (6.18)$$

For example, consider a situation where the water available is 1.2 million m^3 (1000 acre-foot) and the time period of pumping is 1 h (1-time step). Using Eq. (6.18), the equivalent flow is 340 m^3/s.

6.6.2 Generation Mode

Just like in pumping mode, in generation mode, a number of factors should be considered to establish whether the supply of power peaks is required (as compared to baseload supply), and if so a determination should be made on the quantity of energy that is available and/or should be supplied in a timely manner. In general, there are four main limiting factors during generating mode:

1. Supplement of peak power: Is peak power supply required on the power grid? If so, the quantity of energy required should be determined.
2. Unit Capability: What is the capacity of the pump/turbine units (e.g., 200 MW)? And are units capable of operating under single-speed or adjustable-speed conditions?
3. Water Available in Upper Reservoir: What is the quantity of water available in the upper reservoir for turbine operation?
4. Volume Available in Lower Reservoir: This is important since the lower reservoir should be at such a level that it would accept water discharged from the upper reservoir.

Following similar logic with pumping operation, the required flow rate into the turbines can be easily determined (and should take into consideration any time intervals the generation is covering peak loads):

$$E_{turbine} = P_{turbine} \cdot t \qquad (6.19)$$

$$P_{turbine} = \frac{\gamma Q \Delta H}{\eta} \qquad (6.20)$$

Solving for the flow rate, Q,

$$Q = \frac{P_{turbine} \eta}{\gamma \Delta H} \qquad (6.21)$$

For example, consider that the power required is 1,000 MW, with a turbine efficiency of 98%, and net head is 443 m. Using Eq. (6.21), the generated power limiting factor is 235 m^3/s.

Also, in the generation case, the unit capability is calculated based on whether a single-speed or adjustable-speed equipment are used.

In generating mode, the single-speed units can operate over a range of flows down to 50 percent of the rated discharge; however, the range of operation is limited down to 70% to maintain unit efficiency.

As an example, consider a situation where a required power peak supply is 700 MW. A facility has four pump/turbine units with a pump/turbine unit capacity of 325 MW each and turbine efficiency 89%. The net head is 443 m. The solution to find the required flow rate assumes that unit operates when power required is between 70 and 100% of the rated unit capacity to supply the peak load, which results in a turbine range of 228–325 MW: (0.7*325 to 1*325 MW).

The solution first should evaluate Unit 1, and whether peaking power required is greater than 325 MW. Since 700 MW > 325 MW, Eq. (6.21) is used to calculate an equivalent flow for Unit 1 of 84.1 m^3/s. The solution now evaluates Unit 2, and whether the power peak required is greater than 650 MW (2*325 MW = 650 MW), which is that rated capacity of two turbine units. Since 700 MW > 650 MW, Eq. (6.21) is again used to calculate an equivalent flow for Unit 2 of 84.1 m^3/s. The solution next considers Unit 3, and since the expected power required peak is less than 975 MW (3*325 MW = 975 MW), equivalent to operating three units at full capacity, and less than 878 MW (325 MW*2 + 325 MW*0.7 = 878 MW), equivalent to two units operating at full capacity, and the third unit operating at 70%, then Unit 3 and any other subsequent units will not turn on. The units are considered separately, so the flows are added from each of the operating units resulting in a cumulative flow of 168.2 m^3/s.

For variable speed units in generating mode, the turbines are capable of operating at a broader range of power outputs, down to 30 percent of the rated discharge. However, to maintain optimum efficiencies, they are typically operated within 70% of the rated turbine unit capacity. Since adjustable-speed turbines are operated in a similar manner as single-speed ones, the analysis methodology of the deployment scenario used for single-speed turbine units in generating mode also applies. The main advantage in this case is the flexibility offered by each turbine unit in operating at a certain required capacity in order to meet any sudden variations in the demand, and can operate to carry the load (supply at higher than the most efficient 70% capability) until more units are up to speed and capable of sharing the load. This would not be possible in the case of single-speed equipment.

6.7 Case Study: Pumped Hydro Storage in the Hoover Dam (USA) [6]

One of the most recent projects for pumped hydro storage is the augmentation of the Hoover dam, with a pumping system that is expected to turn the dam into a huge battery. The

Los Angeles Department of Water and Power, an original operator of the dam when it was erected in the 1930s, wants to equip it with a $3 billion pipeline and a pumping station powered by solar and wind energy. The pumping station, downstream, would help regulate the water flow through the dam's generators, sending water back to the top to help manage electricity at times of peak demand. The net result would be a kind of energy storage: performing much the same function as the giant lithium–ion batteries being developed to absorb and release power.

The Hoover dam operates the hydroelectric plant that supplies power to Arizona, California, and Nevada. More than 500 feet below the dam is a 2,080 MW power plant operated by the Bureau of Reclamation that. It uses seventeen main Francis turbines or waterwheel located under the generators. Eight of the turbines are on the Nevada side of the plant and nine are on the Arizona side. The last unit was installed in 1961, and an uprate project was completed in 1993. The water reaches the turbines through four penstocks (see Fig. 6.11b).

At a cost of $128 million, the Boulder Dam (renamed Hoover in 1947) was the largest concrete structure in the world. Ice water circulated through more than 582 miles of 1-inch steel cooling pipes installed to help the 4,360,000 cubic yards of concrete to cure properly during the sweltering Nevada build. Using the Colorado River, the dam at the time was the most powerful hydroelectric dam in the world. Currently, it generates 4 billion kilowatt-hours of power per year, enough for 1.3 million people in six states.

The Hoover Dam retrofitting proposal would operate differently. The dam, with its towering 726-foot concrete wall and its 17 power generators that came online in 1936, would not be touched. Instead, engineers propose building a pump station about 20 miles downstream from the main reservoir, Lake Mead, the largest artificial lake in the US. A pipeline would run partly or fully underground, depending on the location ultimately approved.

Los Angeles has two basic motives for this plan. First, the water level of Lake Mead, the nearly 250-square-mile reservoir that provides water to Arizona, California, and Nevada, continues to drop due to long-term drought. The lower water levels are reducing the power that Hoover Dam's electrical turbines generate. Second, California has mandated statewide cuts in fossil-fuel use and increases in renewable energy production.

(a)   **(b)**

(c)   **(d)**

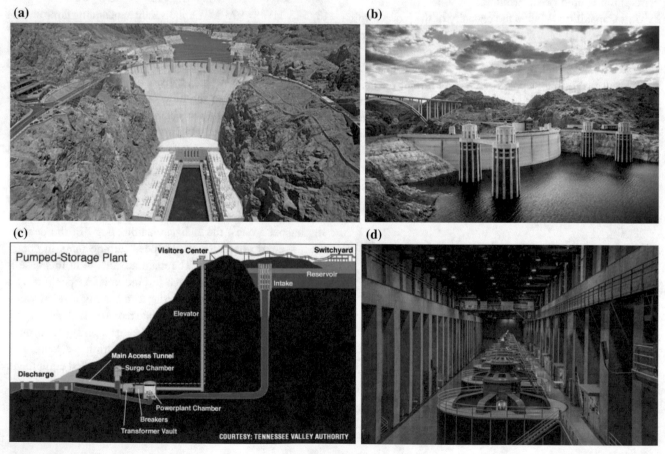

Fig. 6.11 a Hoover Dam showing the upper (lake Mead) and lower reservoirs, **b** The four penstocks leading to the water turbines, **c** Hoover dam site plan and **d** generators room [6]

Many countries including Spain, Norway, Switzerland, and the U.S. already use large pumped hydro storage systems. One key difference here is that this proposed project would use wind and solar electricity to pump the water. As can be seen in Fig. 6.12a–c which shows the sequence of operation of the proposed project, it starts with Copper Mountain Solar Facility providing power to pump the water to Lake Mead (Fig. 6.12a and b), and then the Hoover Dam power station can proceed to generate electricity (Fig. 6.12c), sending the water to the Colorado River (lower reservoir of the Hoover Dam) for the cycle to start again.

(a)

(b)

(c)

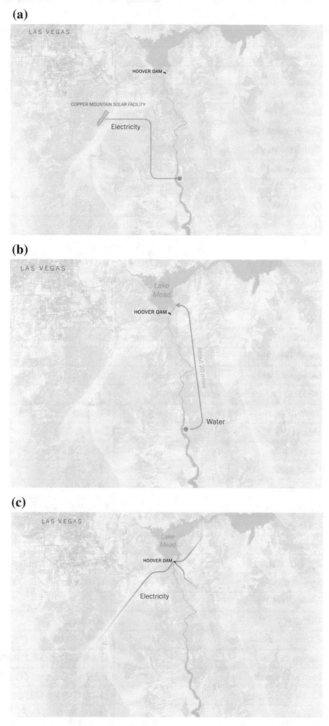

Fig. 6.12 The sequence of operation of the proposed pumping project **a** supplying pumping power to **b** pump the water up the upper reservoir of the dam so it can **c** generate and distribute electricity [6]

The Hoover Dam project may help answer an important question for the energy industry: how to come up with affordable and efficient power storage, which is seen as the key to transforming the industry and helping curb carbon emissions. Because the sun does not always shine, and winds can be inconsistent, power companies look for ways to bank the electricity generated from those sources for use when their output slacks off. Otherwise, they have to fire up fossil-fuel plants to meet periods of high demand.

And when solar and wind farms produce more electricity than consumers need, California utilities have had to find ways to get rid of it—including giving it away to other states—or risk overloading the electric grid and causing blackouts.

Table 6.4a–c shows some facts and figures about the Hoover Dam to better appreciate the magnitude of such a facility and how it contributes to the energy scene in the USA.

Table 6.4 Hoover Dam properties [6]

(a) General facts	
Official Name	Hoover Dam
Location	Clark County, Nevada/ Mohave County, Arizona, US
Status	In use
Construction began	1931
Opening date	1936
Construction cost	$ 45 Million
Owner	United States Government
Operator(s)	U.S. Bureau of Reclamation
Hydraulic head	590 ft (180 m) (Max)
Turbines	17 main, Francis-type
Installed capacity	2080 MW
Annual generation	4.2 billion *KWh*
Nearest city	Boulder City, Nevada
Built	1933
Architect	Six Companies, Inc. (structural), Gordon Kaufmann (exteriors)
Architectural style	Art Deco
Governing body	Bureau of Reclamation
(b) Dam and spillways information	
Type of dam	Concrete gravity-arch
Height	726.4 ft (221.4 m)
Length	1,244 ft (379 m)
Crest width	45 ft (14 m)
Base width	660 ft (200 m)
Volume	3,250,000 cu yd (2,480,000 m^3)
Crest elevation	1,232 ft (376 m)
Impounds	Colorado River
Type of spillway	2 × controlled drum gate
Spillway capacity	400,000 cu ft/s (11,000 m^3/s)
(c) Reservoir information	
Creates	*Lake Mead*
Capacity	28,537,000 acre·ft (35.200 km^3)
Active capacity	15,853,000 acre·ft (19.554 km^3)
Inactive capacity	10,024,000 acre·ft (12.364 km^3)
Catchment area	167,800 sq mi (435,000 km^2)
Surface area	247 sq mi (640 km^2)
Normal elevation	1,219 ft (372 m)
Max. water depth	590 ft (180 m)
Reservoir length	112 mi (180 km)

References

1. M. Richmond, The Niagara Falls hydroelectric station [Online], http://spiff.rit.edu/classes/phys213/lectures/niagara/niagara.html. Accessed 14 Aug 2019
2. H. Ibrahim, A. Ilinca, J. Perron, Energy storage systems—characteristics and comparisons. Renew. Sustain. Energy Rev. **12**(5), 1221–1250 (2008)
3. V. de Turenne, *We are not making this up: Ivanhoe Reservoir in Silver Lake covered in little black plastic balls* (Los Angeles Times, Southern California, 2008)
4. Technical Analysis of Pumped Storage and Integration with Wind Power in the Pacific Northwest Final Report (2009)
5. F. Faure, *Suspension Magnetique Pour Volant D'Inertie* (Jun. 2003)
6. I. Penn, *The $3 Billion Plan to Turn Hoover Dam Into a Giant Battery* (The New York Times, 24 Jul 2018)

7.1 Compressed-Air Energy Storage Systems

The utilization of the potential energy stored in the pressurization of a compressible fluid is at the heart of the compressed-air energy storage (CAES) systems. The mode of operation for installations employing this principle is quite simple. Whenever energy demand is low, a fluid is compressed into a voluminous impermeable cavity, where it is stored under high pressure for the long term, as shown in Fig. 7.1a. For periods where the demand is high, the electrical supply is to be augmented. In this case, the fluid is released from its high-pressure storage and into a rotational energy extraction machine (an air turbine) that would convert the kinetic energy of the fluid into rotational mechanical energy in a wheel that is engaged with an electrical generator and then back into the grid, as shown in Fig. 7.1b.

Just like pumped hydro storage, the large-scale CAES systems benefit from the existence of underground reservoirs that are both cavernous and also impermeable. Depleted natural salt mines, as well as depleted oil and gas fields are perfect candidates for such major storage space requirement, but of course those are not widely available. These reservoirs, however, are key to the pressurized storage of the working fluid, and thus having them available topographically is a natural factor in considering CAES systems as a prominent storage facility. As will be discussed through this chapter, the global efforts for carbon dioxide capture and sequestration can be coupled with CAES systems. Not only would this benefit in facilitating the pumping of oil from wells and also store CO_2 away, it benefits the geology of the oil field after all oil is extracted in order to prevent sinkholes from forming. Also, it would introduce a generalized form of compressed gas energy storage (CGES), which would rely on another gas (CO_2, for example) to be the working fluid instead of air in a closed-loop cycle. It should be mentioned that the energy density of compressed-air systems is lower than that of combustion-based processes, and losses due to airflow are particularly high. However, because the implementation of CAES systems is relatively inexpensive, eco-friendly, mechanically simple, and is easy to maintain, these systems have a great promise. These advantages and more, as more discussion on such systems will unfold in this chapter, explain the reason why air-powered drives could be the answer to the energy storage question.

7.2 Large-Scale CAES Systems

The availability of underground caverns that are both impermeable and also voluminous were the inspiration for large-scale CAES systems. These caverns are originally depleted mines that were once hosts to minerals (salt, oil, gas, water, etc.) and the intrinsic impenetrability of their boundary to fluid penetration highlighted their appeal to be utilized as reservoirs where air can be compressed with no or little leakage out of the system. This guaranteed that the energy stored will not be lost and that the high levels of pressure needed to operate such installations are attainable and sustainable.

For example, the first large CAES installation was the 290 MW plant at Huntorf, Germany, developed by E. ON-Kraftwerk in 1978, as shown in Fig. 7.2a. The plant mitigated any grid supply/demand irregularities by storing electricity as pressurized air during low demand time (at night or weekends) and releasing it again when demand increased as shown in Fig. 7.2b. The plant is still operational and used as a backup power "battery". The compressed air is indeed stored in underground depleted salt caverns that can fill up in 8 h at a rate of 108 kg/s. In discharge mode (supporting the grid during high demand), the compressed air is released and heated up by burning natural gas. The expansion of the air drives a 320 MW turbine for two hours, after which the caverns become depleted (the pressure remaining is not enough to give high quality of energy) and have to be refilled.

The second major CAES plant is run by PowerSouth Energy Cooperative in McIntosh Alabama, as shown in Fig. 7.3a. The 110-megawatt CAES Unit was declared

© Springer Nature Switzerland AG 2020
A. H. Alami, *Mechanical Energy Storage for Renewable and Sustainable Energy Resources*,
Advances in Science, Technology & Innovation, https://doi.org/10.1007/978-3-030-33788-9_7

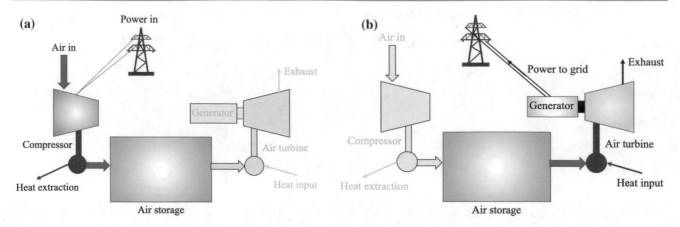

Fig. 7.1 Depiction of operation of a large-scale CAES system in **a** charging mode and **b** discharge mode

commercial in 1991 and is the only one of its kind in the U.S. During off-peak hours, air is pumped into the cavern in a process they label as "compression mode." At full charge, air pressure in the cavern reaches nearly 1,100 lb per square inch (7.5 MPa) as shown in Fig. 7.3b.

During periods of peak, the plant is put into "generation mode." Air from the cavern is released, routed through more than 1,000 feet of pipe and fed into a heat exchanger called a recuperator. Here, it is heated to approximately 600 °F (315 °C). The main difference from the Huntorf facility is that the McIntosh plant allows this hot air to enter a high-pressure combustion chamber, where natural gas is used to further heat the air to around 1,000 °F (537 °C) before entering the high-pressure expander. The exhaust in the high-pressure expander is re-heated to 1,600 °F (871 °C) before entering the low-pressure expander where it is fed back through the recuperator, providing an efficient source of heat for this stage of the process. Excess heat is discharged into the atmosphere at a temperature of around 280 °F (137 °C). Together, th`e high-pressure and low-pressure expanders rotate the generator to produce enough electricity to power nearly 110,000 homes for up to 26 h. Table 7.1 offers a comparison between these two major CAES power plants currently in service.

Remember that the role of the decision-maker is matching the most suitable energy storage technology with the energy resource. For example, wind farms operate around the clock to generate electricity regardless of demand, as the accurate forecasting of demand is far easier than accurately forecasting wind energy availability. This necessitates the usage of a reliable, sustainable, and sizeable energy storage system that would absorb excess generation and then supply the grid back with energy when needed. In this case, CAES systems are a very attractive alternative to chemical or electrochemical storage options. Note that CAES systems can either be charged (air compression into the cavern) mechanically via compressors driven by coupling the

mechanical motion of the wind turbines or by the electrical energy generated from the wind turbines driving electrical motors that operate the compressors as summarized in Fig. 7.4. Of course, the electrical option could be more logical as it would be technically difficult to provide and transmit mechanical energy from the nacelle up the tower of the turbine all the way down to the base of the tower to drive the compressor. The mechanical drive option should still be considered if access to the gearbox on top of the tower is available, since this omits one energy conversion step, and hence will increase the roundtrip storage efficiency.

7.3 Sample Specifications of Components of Large-Scale CAES Systems

There are many practical considerations that arise when operating a large-scale CAES system that may not be too obvious during the abstract design stage. The following lists some of the key components of CAES systems used in the Huntorf of the McIntosh installations that would help the reader form an idea of such considerations.

7.3.1 The Air Compressor

The compressor is a core component of CAES systems, operating at pressure ratios of 40–80 if not more. The Huntorf power plant uses axial flow and centrifugal multistage compression with inter-stage and post-stage cooling (as you probably remember from the power cycle T-s diagrams in thermodynamics, these decrease the required power input and enhance the overall system performance). For smaller CAES systems, it could be more suitable to use a single-stage or multistage reciprocating compressor to reduce the volume of the gas storage device and ensure higher pressure values in storage.

Fig. 7.2 a Image of the
Huntorf CAES facility in
Germany and **b** its principle of
operation [1]

(a)

(b)

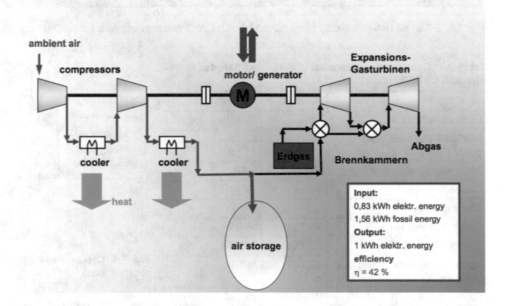

7.3.2   Expander

The sudden depressurization of the stored air entails great
losses, as well as unpredictable behavior of the compressible
gas. And just like with stage-wise compression, it is beneficial to have expansion that takes place also in stages before
reaching the air turbine (or combustion-based gas turbines).
The Huntorf power station uses a modified steam turbine as
its first stage to contend with the expansion of air from high
storage pressures. Small CAES systems would use micro gas
turbine components, reciprocating expanders, or screw air
engines, which are less efficient. These expanders may not
be necessary for small-scale CAES systems.

7.3.3   Air/Gas Storage Vessels

This is the heart of any CAES storage system. Large-scale
CAES usually involves storing gas in depleted salt, water,
oil, or gas fields underground. Smaller scale CAES systems
can use aboveground high-pressure silos or gas storage
containers depending on the selected operational pressures.

7.3.4   Thermal Storage System

Heat exchangers with high effectiveness, ϵ, are required for
the expected issue of heat generated due to air/gas

(a) **(b)**

Fig. 7.3 **a** The PowerSouth energy cooperative McIntosh CAES power plant and **b** the pertinent salt cavern dimensions [2]

Table 7.1 Operational parameters for the Huntorf and McIntosh CAES facilities [3]

	Huntorf	McIntosh
Cycle efficiency (%)	42	54
Maximum electrical input power (MW)	60	50
Maximum air mass rate (kg/s)	108	~90
Charge time (h)	8	38
Discharge time (h)	2	4
Cavern pressure range (bar)	46–72	46–75
Cavern volume (m³)	310,000	538,000
Maximum electrical output power (MW)	321	110
Control range (output) (MW)	100–321	10–110
Setup time (normal/emergency) (min)	14/8	12/7
Maximum mass flow rate (kg/s)	455	154
High-pressure turbine inlet (bar)	41.3 @ 490 °C	42 @ 538 °C
Lower pressure turbine inlet (bar)	12.8 @ 945 °C	15 @ 871 °C
Exhaust gas temperature (°C)	480	370 (before recuperator)

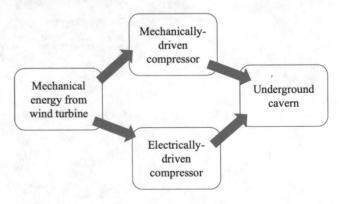

Fig. 7.4 Options for supplying the compressor with work during charging mode

pressurization, and the counter phenomenon of freezing during depressurization, just like the nozzle of a can of aerosol becoming almost frozen after continuous use. From an ideal gas law perspective, we note that there is a correlation between pressure, p, and temperature, T: $PV = nRT$, where n is the number of moles and R is the gas constant. Thus the selected heat exchangers need to be airtight, allow low heat loss, and be adaptable to the working range of the associated turbomachinery, at an economical cost. There are also numerous requirements on the material of the heat storage device. Although all CAES plants operate along the same principle, they have tended to focus on retaining the heat produced during the compression process to use as supply heat during decompression, which will increase the overall efficiency of the plant.

7.4 Classification of CAES Systems

The classification of CAES systems can be done according to their storage capacity, type of application associated with their installation, and according to the operational assumptions on how they handle the heat generated during compression.

7.4.1 Size and Scale

CAES are classified as either large scale and small scale of microscale.

1. Large-scale CAES (>100 MW): These systems are used in conjunction with utilities power grid applications such as load shifting and peak load shaving. Examples of those are the Huntorf and McIntosh power plants.
2. Small-scale CAES (from a few MW to kW): These are mainly used for renewable energy applications, such as

power backup, load tracking, and uninterrupted power supply (UPS) integration.
3. Microscale CAES (~10 kW): These with their limited capacity and storage volume can be used for military applications and as backup power supplies. They can also be found in pneumatic vehicle applications and household power grids.

7.4.2 Heat Generation Handling

The heat generated during air or gas compression produces another classification of CAES systems. Heat transfer reduces the compression efficiency, as the relationship between the power required from the compressor is non-linear, and higher pressures output required from the compressor means that more input horsepower is needed, as shown in Fig. 7.5. This highlights the benefits of using multistage compression rout.

The three main approaches to designing a CAES system are shown in Fig. 7.6, and can be summarized as follows:

1. D-CAES (diabatic) systems: A diabatic process is defined as, "A thermodynamic change of state of a system in which the system exchanges energy with its surroundings by virtue of a temperature difference between them." This assumes that there are no heat collection systems

Fig. 7.5 Power needed from air compressors to achieve storage pressure [4]

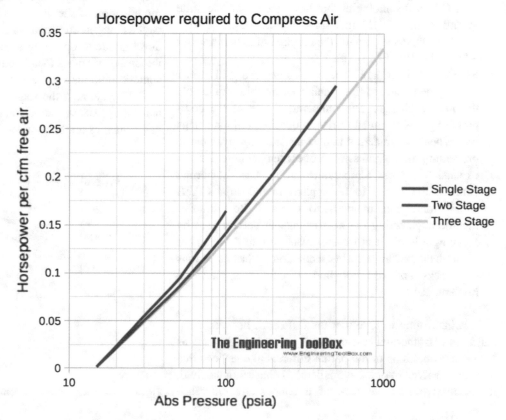

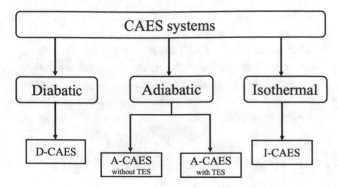

Fig. 7.6 Classification of CAES systems according to how the compression generated heat is handled

(heat exchangers) associated with the facility and thus during compression, heat is transferred directly to the surroundings, thereby causing waste. Such systems will require an external heat source during expansion (decompression) to prevent condensation, or worse, freezing in the air turbine stage. In this case and to provide this external heat, fossil fuel must be burnt with the consequent flue gas emissions and loss of the heat generated during compression.

2. A-CAES (adiabatic) systems: These are the most widely used design approach. The heat generated by compression is transferred and stored in a thermal energy storage (TES) system, which is later utilized during the expansion process. There are also Advanced A-CAES (AA-CAES) technologies that have been available quite recently (since 2015) that uses state-of-the-art ceramic heat exchangers to provide the required high heat transfer efficiencies (see Fig. 4.10).

3. I-CAES (isothermal): These systems are still under development and require specialized machines to handle the heat exchange. The temperature rise of the compressed gas is assumed to rise in quasi-equilibrium steps where heat is transferred almost instantaneously, thereby preventing the compression-process temperature rise and expansion process temperature drop. Quasi-isothermal compression is yet to be applied in industrial CAES installations, and methods to expedite heat transfer include augmenting the heat exchanger surface area by spraying a liquid heat transfer material into the chamber of the heat exchanger. This methodology has its serious drawbacks and did not find a wide-scale industrial implementation.

It is interesting to note that the difference between adiabatic and isothermal assumptions is only apparent as boundary conditions of the partial differential equations that need to be solved in order to predict the behavior of the fluid. It is intuitive that the lack of a temperature difference

between two points (isothermal) can cause no heat transfer (adiabatic), but as boundary conditions they appear as $\Delta T = 0$ and $\dot{Q} = 0$, respectively.

7.5 Governing Equations

For the purposes of research and development of CAES systems, air is usually treated as an ideal gas. However, the behavior of air as a compressible fluid causes nonlinear temperature and pressure interactions. This is more pronounced at the expected operational high temperatures and pressures levels of large-scale CAES systems that are sure to change the behavior of air. The standards developed by Verein Deutscher Ingenieur and the American Society of Mechanical Engineers (ASME) illustrate the hypothetical limit for when to consider a gas to behave as an ideal one. Thermodynamic properties such as heat capacity, volume, enthalpy, and entropy for humid gases at different temperatures and pressures of ideal and real gases at different temperatures and pressures are different according to the exergies at those conditions (shown in Fig. 7.7), and it can be concluded that the exergies for real and ideal gases are different.

For better accuracy of the design of CAES systems, a better equation set has to be adopted in order to model the behavior of the working fluid. Gas dynamics equations present a good approximation as they utilize the isentropic modification of the ideal gas law, which would work very well as long as shockwaves are avoided. These shockwaves develop due to operating a compressible fluid at or beyond the speed of sound (Mach number > 1). Even these supersonic conditions are accommodated within the isentropic model by defining the stagnation (total properties) and critical (sonic) states, the details of which can be found in any gas dynamics textbook. Here, the focus will be only on the

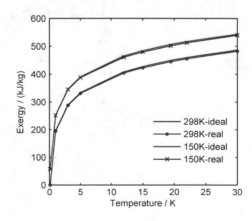

Fig. 7.7 Exergies versus temperature for ideal and real gases [4]

development and implementation of the isentropic relations that will assist in the CAES system design without reaching the critical or sonic states.

The isentropic fluid flow model assumes that it is both adiabatic and reversible. That is, no heat is added to the flow, and no energy losses occurring due to friction or dissipative effects. In reality, the heat addition due to compression is handled by the heat exchangers. For an isentropic flow of a perfect gas, several relations can be derived to define the pressure, density, and temperature along a streamline. After lengthy derivations of the differential energy equations, the isentropic model of ideal gas takes the following form [5]:

$$PV^\gamma = \text{constant} \qquad (7.1)$$

where γ is the heat capacity ratio. This can be written as

$$\frac{P_2}{P_1} = \left(\frac{V_1}{V_2}\right)^\gamma \qquad (7.2)$$

This equation can also be used to estimate the temperature (by substituting the equation of state for an ideal gas, $PV = nRT$):

$$TV^{\gamma-1} = \text{constant} \qquad (7.3)$$

The same can be applied for the total volume and density:

$$\frac{P_2}{P_1} = \left(\frac{T_2}{T_1}\right)^{\frac{\gamma}{\gamma-1}} = \left(\frac{V_1}{V_2}\right)^\gamma = \left(\frac{\rho_2}{\rho_1}\right)^\gamma \qquad (7.4)$$

The value of γ is constant for a given gas. The values of γ for some gases are given in Table 7.2.

Equations (7.1)–(7.4) can also be written in terms of the Mach number, M. Remember that the Mach number is the ratio of the flow velocity, v, over the speed of sound, a, within the fluid: ($M = \frac{v}{a}$). The speed of sound is defined as

$$a = \sqrt{\gamma \frac{P}{\rho}} = \sqrt{\gamma R T} \qquad (7.5)$$

For example, the pressure ratio as a function of the Mach number becomes

$$\frac{P}{P_{\text{critical}}} = \left(1 + [(\gamma - 1)/2]M^2\right)^{-\gamma/(\gamma-1)} \qquad (7.6)$$

Note that the sonic conditions apply for the sonic conditions and hence P_{critical} is assumed as P_2. The critical area, A^*, is the minimum cross-sectional area of the expansion nozzle and it can also be calculated as

$$\frac{A}{A^*} = \left(\frac{\gamma+1}{2}\right)^{-\frac{\gamma+1}{2(\gamma-1)}} \frac{\left(1 + \frac{\gamma-1}{2}M^2\right)^{\frac{\gamma+1}{2(\gamma-1)}}}{M} \qquad (7.7)$$

Table 7.2 Values of specific heat ratio for some common gases [5]

Gas	Specific heat ratio, γ
Acetylene	1.30
Air, standard	1.40
Ammonia	1.32
Argon	1.66
Carbon dioxide	1.28
Carbon disulphide	1.21
Carbon monoxide	1.40
Chlorine	1.33
Helium	1.66
Hydrogen	1.41
Methane	1.32
Natural gas (methane)	1.32
Nitrogen	1.40
Oxygen	1.40
Propane	1.13
R-12	1.14
R-22	1.18
R-134a	1.20
Steam (water)	1.33

The conversion of potential energy (pressure in the cavern) into kinetic energy in the nozzle leading to the air turbine can be analyzed by employing an isentropic assumption to govern the expansion process. Since the flow is assumed to remain subsonic once decompression takes place, the critical pressure ratio between the inlet and outlet will have to be higher than 0.5283 to ensure no shockwaves will occur at the nozzle (chocking the nozzle). This makes the exit conditions known in terms of the geometry. And since velocities within the air reservoir are much smaller than those at the outlet, the former are assumed to be zero and thus stagnation (total) conditions prevail at the inlet. For isentropic expansion, we can start with the simple principle of gas dynamics:

$$\frac{p_{t2}}{p_{t1}} = \frac{p_2}{p_1}\left(\frac{1 + [(\gamma-1)/2]M_2^2}{1 + [(\gamma-1)/2]M_1^2}\right)^{\gamma/(\gamma-1)} \qquad (7.8)$$

For stagnation conditions at the inlet, we have $p_1 = p_{t1}$ and $M_1 = 0$ and the equation becomes (see Fig. 7.8a)

$$\frac{p_2}{p_{t2}} = \left(\frac{1}{1 + [(\gamma-1)/2]M_2^2}\right)^{\gamma/(\gamma-1)} \qquad (7.9)$$

This is a function of the Mach number at the exit and can be plotted for the pressure ratio p_2/p_1 remembering that $p_1 = p_{t1} = p_{t2}$, as shown in Fig. 7.8b.

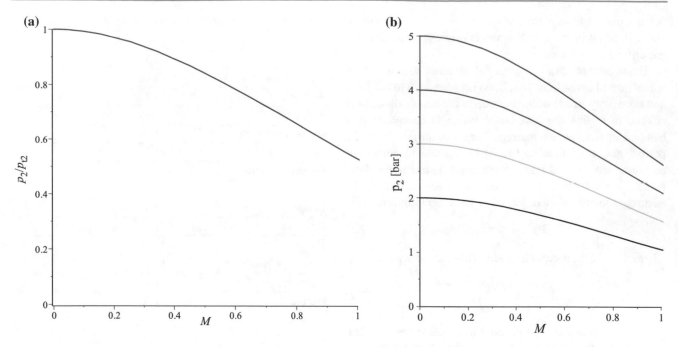

Fig. 7.8 **a** Exit stagnation pressure ratio variation with outlet Mach number and **b** pressure ratio variation with the outlet Mach number [6]

Figures 7.8 enables the estimation of the maximum exit velocity and various outlet pressures by fixing the starting pressure at a certain preset value. Thus, the maximum theoretical velocities issuing from the nozzle at this area ratio can be used later to obtain the efficiency of the storage system during the discharge part of the cycle.

The definition of the critical state makes fixing static states easier, since the absolute temperature between initial state 1 and final state 2 is assumed to be constant ($T_1 = T_2 = T$). The work done on a control volume of air taken within the compression cylinder between these two states is given by

$$W_{1 \to 2} = \int_{V_2}^{V_1} p dV = \int_{V_2}^{V_1} \frac{nRT}{V} dV = nRT \int_{V_2}^{V_1} \frac{1}{V} dV \quad (7.10)$$

This can be expanded by carrying out the integration:

$$W_{1 \to 2} = nRT(\ln V_1 - \ln V_2) = nRT \ln \frac{V_1}{V_2} \quad (7.11)$$

In a reservoir, it is easier to measure pressures than to estimate the gas volume, and Eq. (7.11) can also be written more conveniently in terms of starting and terminal pressures P_1 and P_2, respectively, and the volume of the storage reservoir,

$$W_{1 \to 2} = P_1 V_1 \ln \frac{P_1}{P_2} \quad (7.12)$$

Since adiabatic conditions are assumed to hold, a simplified form of the energy equations is applied between points 1 and 2 to estimate the maximum exit velocity of air from the nozzle to be

$$v_{\max} = \sqrt{2 \frac{P_1}{\rho_{ave}}} \quad (7.13)$$

where ρ_{ave} is the average air density between states 1 and 2, assuming a linear pressure drop during system discharge. The overall CAES system efficiency as a function of the pressure ratio P^* is given as

$$\eta = \frac{\ln P^* - 1 + \frac{1}{P^*}}{\frac{P^{*\frac{(n-1)}{n}} - 1}{n-1} + P^{*-\frac{1}{n}} - 1 + (P^* - 1)\left(P^{*-\frac{1}{n}} - \frac{1}{P^*}\right)} \quad (7.14)$$

The angular velocity of the air turbine, ω in rad/s, can be estimated from the dimensions of the air turbine wheel based on the velocity of the air stream leaving the expansion nozzle. It can later be used in the basic torque equation $T = I \times \alpha$, where I is the moment of inertia of the turbine rotor and its blades and α is the uniform angular acceleration in rad/s². Starting from rest, the angular acceleration of the wheel is estimated to be the rate of change of the angular velocity over rotation time. Losses in the form of friction, expansion, fitting, windage, and compressibility are responsible for any discrepancies between the calculated velocities and the measured ones.

The electrical conversion efficiency of the system will be taken as the ratio of the electrical output power over the stored pressure potential (P times the total volume of the cylinder) as

$$\eta_{\text{electrical}} = \frac{i \cdot v \cdot t}{P \cdot V} \tag{7.15}$$

where i is the RMS current, v is the RMS voltage, and t is the discharge time.

Finally, the mechanical efficiency of the system is defined as

$$\eta_{\text{mech}} = \frac{\frac{1}{2} I \omega^2}{P \cdot V} \tag{7.16}$$

where I is the moment of inertia of the generator rotor, and ω is its angular velocity in rad/s.

7.6 Modular and Small-Scale CAES Systems

One of the bottlenecks facing the widespread utilization of CAES systems is the availability of a suitable storage reservoir. Storage caverns large enough to ensure the economic and technical feasibility of CAES systems' implementation are scarce and may not exist in a location that is geologically suitable and at the same time rich in storable energy. A solution for this issue resides in following a cellular approach to storage, where modular cylinders can be interconnected and discharged either in tandem or in sequence depending on the energy needs of the grid as depicted in Fig. 7.9.

Not only are the size (volume) and number of cylinders determining factors of the storage capacity of the storage system, they also provide the existence of more than one cylinder that is advantageous in controlling the discharge of the stored compressed air. It makes the system responsive in both high energy density requirements (sequential operation

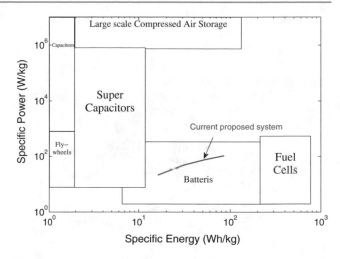

Fig. 7.10 Ragone plot showing the operational range of the modular CAES system [6]

of cylinders) and high power density requirements (tandem operation of cylinders). The Ragone plot of the system is shown in Fig. 7.10, where the overlap with chemical batteries is shown. Large-scale CAES systems offer higher power densities since they operate at much higher storage pressure values (around 70 bar). Another advantage of the modular design is the handling of the heat generated due to compression.

7.6.1 System Design

The experimental unit that is designed for the assessment of the feasibility of such system (shown in Fig. 7.11) is made up of three 7-liter steel cylinders, with air pressure provided by a two-cylinder reciprocating air compressor (PowerPlus-POWX1730) with a maximum flow rate of 180 l/min and maximum working pressure of 10 bar is used. The cylinders will be charged at pressure values between 2 and 5 bar (gauge) for two reasons. First, low pressures would result in small temperature rise within the cylinders, and thus no significant amounts of heat are expected to result from the compression cycle. Second, it is far safer to use low pressures than high pressures. Another added advantage is that once the low end of pressure values is explored, higher levels of energy quality can be expected when operating at higher pressures, and thus any conclusion drawn from the current system would be a conservative one.

Experimentally, the cylinders are connected in parallel via 1/4-inch-diameter PVC pipe network and brass fittings. To better control the charge cycle, three manual air valves (10 bar max) are installed at the inlet of each cylinder, with a one-way valve similarly installed at the cylinder inlet to prevent the backflow of air. To monitor the pressure of the compressed air within the cylinders, three analog pressure

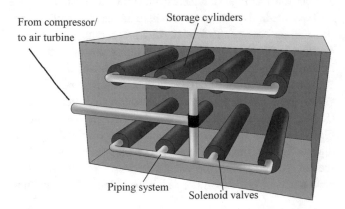

Fig. 7.9 Modular CAES system with cellular storage [6]

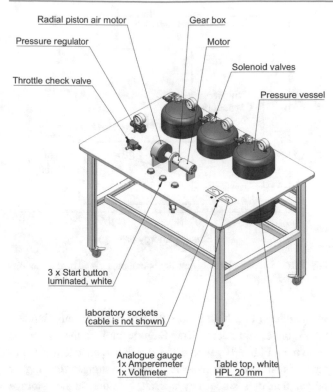

Fig. 7.11 The experimental setup of the modular CAES system [6]

gauges (7 bar max) are attached to the side of each cylinder. The discharge cycle of the compressed air in the cylinders is controlled via a PLC (Siemens LOGO 230 RC, 220 AC, 8 inputs) directing three 5/2-way solenoid valves (4V210-08, 220 AC input, 1.5–8 bar operational pressure) that would either open in unison to give maximum power density, or in sequence to provide maximum energy density. The air turbine/electrical generator assembly shown in Fig. 7.12 is manufactured by Hüttinger (http://www.huettinger.de). The air turbine handles a maximum torque of 800 N cm, while the electrical generator produced 24 V and 33 A at a maximum rotational speed of 3350 rpm.

Fig. 7.12 The air turbine with the coupled electrical generator

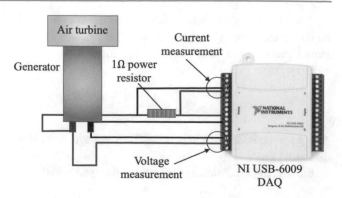

Fig. 7.13 Electrical wiring diagram of the data acquisition system

The electrical output (current and voltage) from the generator is recorded by an NI 6009 USB DAQ module at a 500 Hz sampling rate. The current is measured through a connection to a 1 Ω shunt resistor, R (current = voltage/R). The unit is connected to a computer via USB connection and is used to log the data as shown in Fig. 7.12. (Fig. 7.13).

A tachometer is used to record the rotations of the turbine shaft and compare it with the discharged air velocity experimentally.

The charging time for each cylinder to a pressure of 5 bar is 40 s. The charging side of the cycle is not shown in results since we assume that the availability of air during low demand times from the wind turbine generators, and hence we place no constraints on input air availability for charging.

7.6.2　System Operation

The discharge time determination is of significance in predicting how long the storage system can operate while providing high-quality energy. This is especially important in the sequential operation of the system, where the programming of the PLC controlling the discharge requires the explicit knowledge of when to open/close the solenoids and ensures smooth operation and transition. Estimating the required discharge time entails the processing of raw voltage and current readings obtained from the DAQ system.

Two methods are proposed here to calculate the time constant for each of the four starting pressure ratings of 2, 3, 4, and 5 bar (gauge) at which the resulting power values can be statistically processed to extract a representation of the equivalent direct current (DC) values. The first method fits voltage readings to a polynomial, and the root mean square (RMS) values are calculated correspondingly as shown in Fig. 7.14. The time constant is then determined by the intersection of the RMS value and that polynomial. The second method allows the voltage to drop down to a chosen percentage of the maximum value, and time is read at that

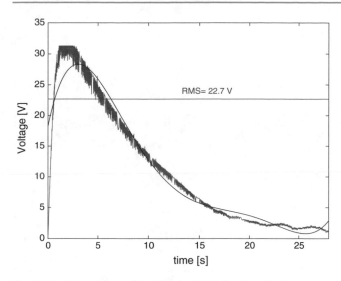

Fig. 7.14 Time constant determination from the plot of voltage versus discharge time for one cylinder with starting pressure of 3 bar [6]

Table 7.3 Time constants results for two RMS and 80% cutoff for different pressures [6]

Pressure (bar)	Time (s)	
	RMS	80% max. voltage
3	5.36	6.31
4	6.01	6.851
5	7.32	7.92

point. A conservative 80% cutoff is chosen here, eliminating the need to recharge from static conditions at each trial and minimizing noise. The calculated time constants from both methods are tabulated and compared in Table 7.3.

The next step is to evaluate the system during sequential (energy density) and tandem (power density) operation.

7.6.3 Sequential Operation of the Three Cylinders from Different Starting Pressures

To obtain a system with higher energy density (longer discharge time at the cost of maximum power), a three-cylinder setup controlled by a PLC to discharge air sequentially with no time delay is used and is shown to almost triple the discharge time compared to a single-cylinder discharges (shown in Fig. 7.14). The resulting voltage with respect to discharge time is shown in Fig. 7.15a–c for the cases of 3, 4, and 5 bar, respectively. The figures show three quasi-identical output signals produced an RMS voltage value of 25, 32, and 31.8 V for discharge times of 20, 21, and 24 s, for the respective ascending starting pressure

values. The real-time voltage fluctuations are due to the large number of readings (500 Hz), and hence a polynomial fit (shown in black) can be used along with the RMS values for better understanding of the system operation.

7.6.4 Tandem (Simultaneous) Discharge of the Three Cylinders from Different Starting Pressures

To obtain high power density, the three cylinders are discharged in unison using the same PLC to control the timing of discharge valve opening. It is observed that this setup can considerably increase the maximum output voltage and current fit for high power density applications. Also, the discharge time for 5 bar starting pressure is seen to reach a little over 96 s compared with the 37 s it took for the discharge of one at the same pressure. The results of the discharge voltage with time are shown in Fig. 7.16a–c for the 3, 4, and 5 bar cases, where the RMS voltage is observed to be 22.7, 28.5, and 31.6 V, respectively.

The three-cylinder setup resides in a location of the Ragone plot (see Fig. 7.10) that is attractive from both energy density and power density vantage points. And since the set discharge times are low, the system can deliver large amounts of power when required by the simultaneous three-unit cylinder discharge. Other discharge combinations can also be programmed into the PLC control unit to accommodate any change in demand, or when more energy than power is needed. Since the system is modular in design, adding more cylinders to the setup would extend the active range of the system to reach the large-scale CAES systems, or by operating at higher pressure ratings (pressure values over 5 bar). In either case, larger losses are expected due to compression, heat, friction, and fittings. These losses cannot be neglected for the large-scale system.

7.6.5 System Efficiency

The efficiency of the system consists of the conversion efficiency of pressure potential energy within the cylinders into kinetic energy within the discharged air, and also the mechanical efficiency of the air turbines handling the ultimate energy conversion into electricity. By comparing the theoretical and experimental air exit velocities as shown in Fig. 7.17, a good correlation of the two trends is observed since no shockwaves were observed experimentally.

Table 7.4 also lists the calculated mechanical and overall efficiencies for different starting pressures according to Eqs. (7.15) and (7.16) for the three-cylinder (energy density) setup:

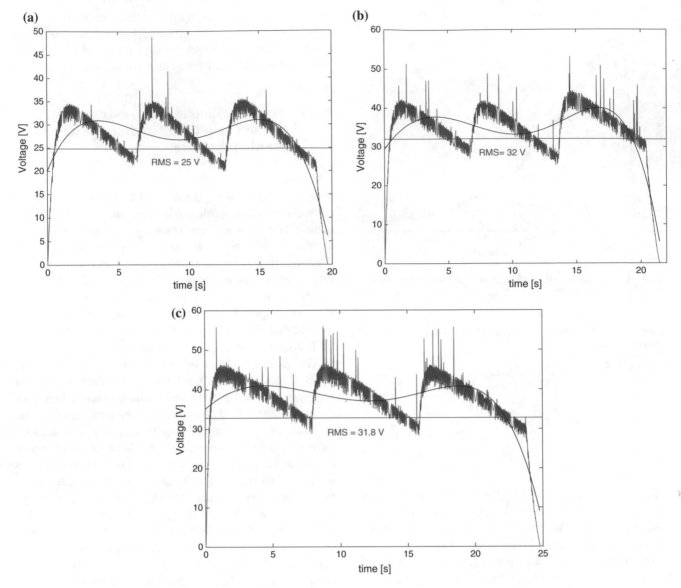

Fig. 7.15 Voltage versus discharge time for three cylinders discharged sequentially from a starting pressure of **a** 3 bar, **b** 4 bar, and **c** 5 bar [6]

7.6.6 Comparison with Batteries Storage

The output RMS voltage and current of the different permutations of the compressed-air storage system are shown in Table 7.5, which compares different setups to batteries with different capacities to put the suggested air storage system in perspective.

As an example to illustrate the results in the table, if 57 cylinders (height × diameter = 408 × 150 mm) operating at 5 bar are interconnected, they would cover a volume of around 0.6 m³, with cylinders operating fulfilling the job of four 24 V batteries for 20 consecutive hours (the benchmark battery is a Trojan U1-AGM, http://www.trojanbattery.com/product/u1-agm/). This is a very small footprint for a

stationary storage system operating at the low-pressure value of 5 bar and offering the flexibility over batteries for either having power density or energy density within the same storage system. To put matters into better perspective in terms of system weight, each cylinder is 1.2 kg, while each of the batteries weighs 12 kg. Thus, the total weight of 57 cylinders is 30% higher if we decide to include the weight of the steel fire extinguishers instead of the actual mass of air (the active material). On the other hand, if system weight is an issue, then the air cylinders can be made from aluminum or less thick steel since the operational pressures will not exceed 5 bar. This might also be reflected in the cost of these systems as the four batteries cost around $400 (for maximum of 2 years), while 57 cylinders will cost around $712 (almost

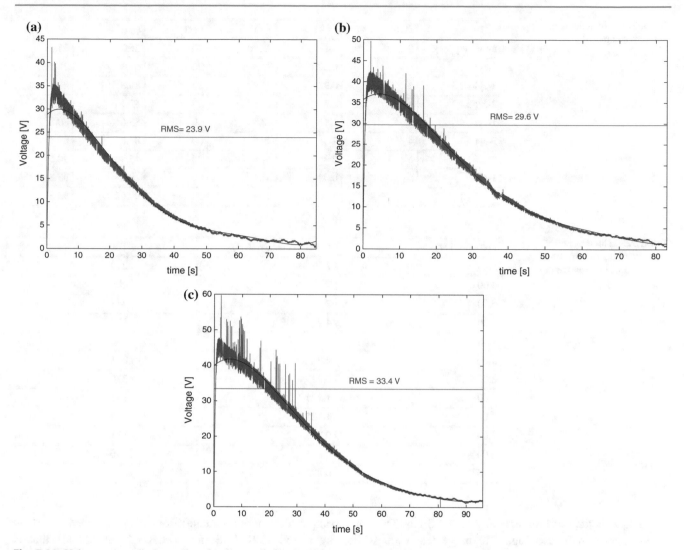

Fig. 7.16 Voltage versus discharge time for three cylinders in tandem charged initially at **a** 3 bar, **b** 4 bar, and **c** 5 bar [6]

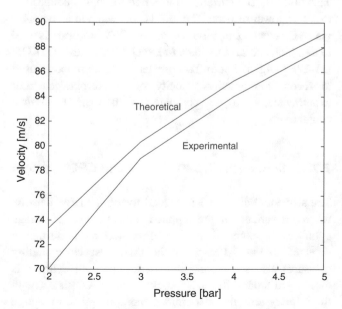

Fig. 7.17 Theoretical versus experimental air exit velocity [6]

infinite number of charge/discharge). Thus, the break-even point with a battery system is around 2 years of continuous operation.

7.7 Compressed Gas Energy Storage (CGES) Systems

Air is not the only option for a working fluid of a modular systems. With the importance of progress in carbon dioxide capture and sequestration, the existence of CO_2 storage facilities has prompted a plan to tap into the stored gas at high pressures, pass it through a power cycle to convert the potential energy into kinetic, then mechanical, and eventually electrical. Another modular low-pressure compressed gas energy storage system will be examined. The system is a closed-loop one, drawing carbon dioxide potentially from underground caverns into a number of pressurized cylinders

Table 7.4 Efficiencies of different setups [6]

	P (bar)	I (A)	V (V)	t (s)	$\eta_{overall}$	η_{mech}
3-cyl-seq	3	1.1	25	20	26.2	61.5
1-cyl	3	0.62	22.7	90	60.3	78.5
3-cyl-all	3	0.92	23.9	80	83.8	80.3
3-cyl-seq	4	1.6	32	23	42.1	66.7
1-cyl	4	1.1	28.5	30	33.6	80.1
3-cyl-all	4	1	29.6	80	84.6	89.4
3-cyl-seq	5	2	31.8	25	45.4	70.5
1-cyl	5	0.92	31.4	90	79.0	93.2
3-cyl-all	5	1.1	33.4	93	97.62	95.6

Table 7.5 Three-cylinder (sequential operation) setup compared to various battery capacities [6]

Pressure (bar)	Battery capacity (AH)	Total
3.00	7	210
4.00	7	699
5.00	7	453
3.00	4.2	839
4.00	4.2	420
5.00	4.2	273
3.00	1.125	226
4.00	1.125	113
5.00	1.125	74
3.00	0.875	181
4.00	0.875	90
5.00	0.875	57

where CO_2 is kept at pressures 2, 2.5, and 3 bar. The minimalist approach is used again to prove that even while operating at such low levels of starting pressures, the system can still produce results that are comparable to battery storage and are scalable. The cylinders exhaust into an air turbine/generator system that allows the conversion of the potential energy into electrical energy. The discharge from the cylinders happens either sequentially for applications requiring high energy density or in unison for high power density applications. The same test rig used for the modular CAES case has been used to identify important experimental parameters, a carbon dioxide source, pressure regulator, and heater to control the pressure and temperature, respectively, of the carbon dioxide entering the cylinders. The operation at the specified pressure range is consistent with the adiabatic assumptions and expected ideal gas behavior of the carbon dioxide. Voltage and current measurements are made for each discharge case, and the energy, power, and overall system efficiency are calculated for each case and compared to similar compressed-air energy storage (CAES) systems. A schematic of the test setup is shown in Fig. 7.18. The only difference for this setup compared to the one described for modular CAES system is the data acquisition. The electrical

output (current and voltage) from the generator is recorded by an NI 9221 C Series Voltage Input Module that is attached to an NI cRIO-9025, which records a measurement each 500 ms. The current is measured through a connection to a 1 Ω power resistor. The unit is connected to a computer via an Ethernet connection where LabVIEW software is used to log, plot, and save the data to Excel documents. LUTRON DT-2259 Digital Photo Tachometer w/Stroboscope is used to record the angular velocity of the turbine/generator experimentally to be compared with their theoretical counterparts.

7.7.1 Sequential Operation of the CGES System

The sequential discharge is needed for high energy density, where the variation in the supplied voltage out of the storage cylinders is expected to be constant. The performance depicted in Fig. 7.19a–c for the three respective starting pressures also show consistent trends, as each cylinder is discharged around the 3.2 s mark, which is consistent with the primary tests that determined the appropriate discharge time that preserves energy quality. The energy density

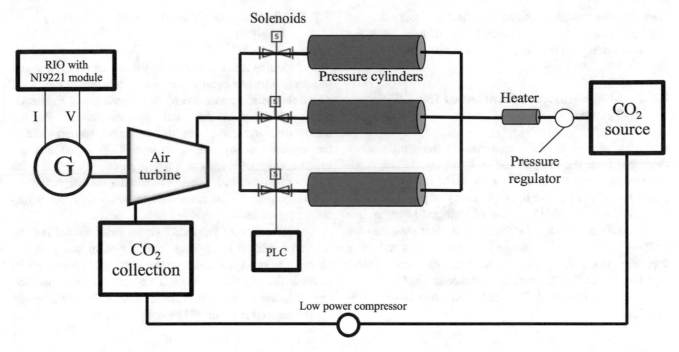

Fig. 7.18 Schematic of the modular CGES system [7]

Fig. 7.19 Sequential discharge for **a** 3 bar, **b** 2.5 bar, and **c** 2 bar [7]

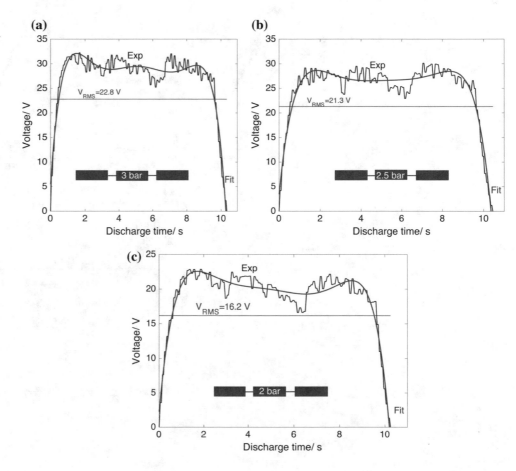

discharge time is programmed via the PLC controller and can be varied to provide various levels of total discharge times depending on the desired application.

7.7.2 Simultaneous Operation of the CGES System

For high power density requirements, the simultaneous discharge from the three cylinders is deployed. The maximum voltage values cannot be maintained for long periods of discharge times due to the rapid decay of pressure within the cylinders. The RMS values of voltage, however, are sustained for up to 100 s for the 3 bar starting pressure case at 22.4 V, which is far more than the duty cycle of a supercapacitor. The voltage variation with time shows identical trends for the three simultaneous discharge from the three cylinders at 3, 2.5, and 2 bar, and is shown in Fig. 7.20a–c.

7.7.3 Kinetic Energy and System Efficiency Analysis

The air turbine rotational velocity for the three experimental protocols is plotted against the starting pressure in Fig. 7.21. The tachometer records the highest, lowest, and final rotational velocity value for each experimental run. For the current discharge conditions, the sequential discharge shows the highest velocity values (around 4750 RPM) and the smallest range of variation. There is an overlap between the single-cylinder discharge and all-cylinder discharge, but the advantage goes to the latter case, especially at higher discharge pressures.

Since the system flexibility stems from controlling the discharge time, it is expected that imposing less stringent energy quality requirements on the sequential operation will increase the overlap between it and all-cylinder operation region. In any case, the maximum achievable kinetic energy for the maximum rotational velocity is around 125 kJ/kg of

Fig. 7.20 Sequential discharge for **a** 3 bar, **b** 2.5 bar, and **c** 2 bar [7]

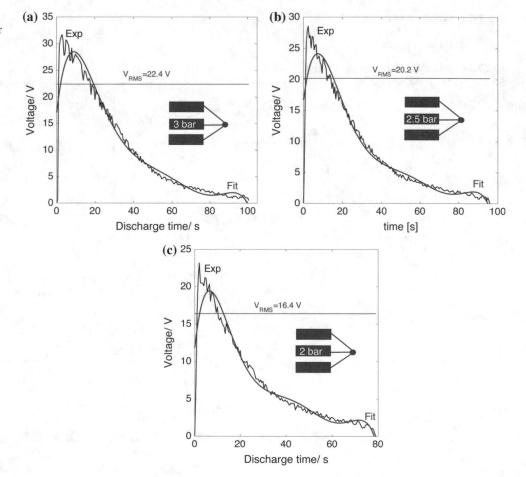

Fig. 7.21 Air turbine rotational velocity [7]

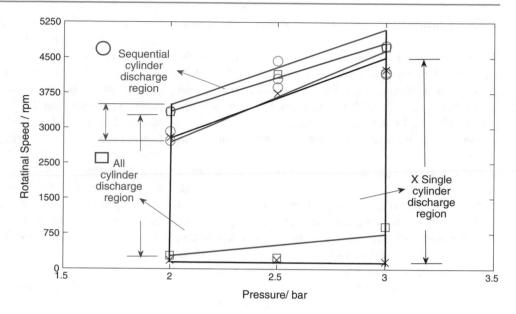

Table 7.6 Summary of overall system efficiency at various operating conditions [7]

Operation mode	Single cylinder			All cylinders			Sequential		
Initial pressure (bar)	2	2.5	3	2	2.5	3	2	2.5	3
Efficiency (%)	24.0	48.6	79	35.2	61.2	76.4	18	36.8	46.2

the rotor mass, accredited to the high-performance air turbine used for experiments.

The overall efficiency of the system for various operating conditions is summarized in Table 7.6.

The overall efficiency is highest for the complete discharge of a single cylinder at 3 bar starting pressure, followed closely by the simultaneous discharge of all cylinders at 3 bar. The higher energy quality at 3 bar for all operating modes manifests itself with the reported higher efficiency values. It is important to note that the low power compressor is expected to consume some of the power during periods of low demand to create the closed-loop operation.

Some available piston-type, 1/8 hp (from Grex USA that operates at 20 L/min and produces an outlet pressure of 4 bar) compressor will operate over an extended period of time (6–8 h) to evacuate the CO_2 collection container shown in Fig. 7.18, and thus is not expected to operate at its full power. Also, the compressor will not overlap with the charge/discharge operation of the cylinders and thus will not affect the overall efficiency of the system.

7.7.4 CGES Ragone Plots

To compare the system response under various load requirements, the Ragone plot is produced and shown in Fig. 7.22 for the sequential operation as well as for the simultaneous discharge of all cylinders.

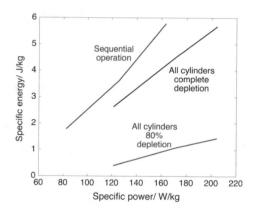

Fig. 7.22 Ragone plot for sequential and two cases of simultaneous discharge: complete and up to 80% depletion [7]

Since the sequential discharge is meant to provide better energy coverage, it reigns this range in the Ragone plot. The complete discharge of all cylinders shows the expected higher specific power output while being unable to cover ranges lower than 120 W/kg. It is interesting to note that if the cylinders are not completely discharged, their performance in terms of energy density is completely restricted, while still capable of providing the same level of power density as when completely discharged. The only advantage of the latter case is shorter charging times, especially when operating with multiple cylinder sets (multiples of three), where depleted units are charged as the operational ones are discharging if the needs arise.

7.7.5   Comparison with Other Storage Methods

The widely used lithium–ion batteries are regarded as the most electrical storage devices. The current gas storage system has a gravimetric energy of 0.3–1.5 Wh/kg and power density of 120–220 W/kg. Table 7.7 shows a comparison with the Li–ion batteries and supercapacitor storage technologies (see https://www.supercaptech.com/battery-vs-supercapacitor).

The compressed gas energy storage system stands out in terms of cost, safety, and cyclability. Also, the chemical, thermal, and electrical stability of the system makes it a natural contender for traditional storage technologies, especially when directly coupled with a charging mechanism that used excess mechanical energy, for example, from a remote wind farm during low demand times. Compared to the similar CAES system developed by our group and detailed in reference, the 3 bar operation of using carbon dioxide as a working fluid significantly reduces the number of equivalent cylinders needed to be equivalent to energy storage in

batteries by 47–61%. Table 7.8 shows the number of 7-liter cylinders (ϕ = 15 cm and L = 40 cm).

The 7-liter cylinder is 15 cm diameter and 45 cm length. This means that 100 cylinders require less than a meter cube (0.8 m^3) of storage space for the 24 h operation of various capacity batteries. The 7 AH battery is capable of operating compressors for air conditioning and refrigerating units requires around 7 m^3 of storage space, which can easily be housed in a small portion of a standard storage room measuring less than 2 × 3 × 3 m in dimensions. The number of cylinders can also be reduced using bigger cylinders or operating at higher pressures. The latter is not recommended in the case of CGES system to minimize the heat transfer issues of cyclic system operation.

7.7.6   Advantages and Disadvantages

In general, the application of CO_2 as a working fluid has enabled the operation at lower starting pressures and required

Table 7.7 Summary of overall system efficiency at various operating conditions [7]

Feature	Li–ion battery	Supercapacitor	CGES
Gravimetric energy (Wh/kg)	100–265	4–10	0.3–1.5
Volumetric energy (Wh/L)	220–400	4–14	>7
Power density (W/kg)	1,500	3,000–40,000	121–220
Voltage of a cell (V)	3.6	2.7–3	16.2–22.8
ESR (mΩ)	500	40–300	NA
Max. efficiency (%)	75–90	98	79
Cyclability (nb recharges)	500-1,000	500,000	500,000
Life	5–10 years	10–15 years	>20 years
Self-discharge (% per month)	2	40–50 (descending)	10%
Charge temperature (°C)	0–45 °C	−40 to 65 °C	20 °C
Discharge temperature (°C)	−20 to 60 °C	−40 to 65 °C	20 °C
Deep discharge pb	Yes	No	Yes
Overload pb	Yes	No	No
Risk of thermal runaway	Yes	No	No
Risk of explosion	Yes	No	No
Charging 1 cell	Complex	Easy	Easy
Charging cells in series	Complex	Complex	Easy
Voltage on discharge	Stable	Decreasing	Stable
Cost per kWh	200–1,000 €	10,000 €	200–400 €

Table 7.8 Equivalent cylinders required for the operation of one battery [7]

Pressure (bar)	Battery capacity (AH)	Total cylinders required		
		CAES	CGES	Cylinders less (%)
3.00	7	1483	900	61
3.00	4.2	839	542	65
3.00	1.12	226	109	48
3.00	0.875	181	86	47

Table 7.9 Advantages and disadvantages of utilizing carbon dioxide as a working fluid [7]

Advantage	Disadvantage
Offers both power and energy density options for small-scale operation	Setup, pressure accumulation, programming require a long lead time
CO_2 density higher than air allowing for higher energy output at lower pressures	Higher pressures than 3 bar will require heat compensation
CO_2 being free of moisture	System maintenance and troubleshooting can be challenging
Can be automated with minimal human interaction during operation	Individual pressure regulators are required per cylinder
One compressor could be used to fill up the tanks. Easy to fill tanks in series	Individual solenoids are needed per cylinder
Generators are added in parallel for higher output. Power that are able to handle starting surge	Cost of modular air generator is high
Larger cylinders could be used as lower pressures are used for maximum power output	–
Layout could be customized per energy requirement. Valves could be added to deter compressed gas to main? Modular systems	–
Low cost per kWh compared to other electrochemical storage devices	–
High product lifecycle with comparable efficiencies with existing electrochemical storage devices	–

a smaller number of storage cylinders than its compressed-air counterpart. There are also some other important points of comparison that are highlighted in Table 7.9.

References

1. D.-I. F. Crotogino, Einsatz von Druckluftspeicher-Gasturbinen-Kraftwerken beim Ausgleich fluktuierender Windenergie-Produktion mit aktuellem Strombedarf 1, www.kbbnet.de. Accessed 14 Sep 2019
2. Compressed Air Energy Storage, http://www.powersouth.com/wp-content/uploads/2017/07/CAES-Brochure-FINAL.pdf. Accessed 19 Aug 2019
3. Q. Yu, Q. Wang, X. Tan, G. Fang, J. Meng, A review of compressed-air energy storage. J. Renew. Sustain. Energy **11**(4), 042702 (2019)
4. Horsepower required to Compress Air Online air compressor horsepower calculator, https://www.engineeringtoolbox.com/horsepower-compressed-air-d_1363.html. Accessed 22 Aug 2019
5. R.D. Zucker, O. Biblarz, *Fundamentals of Gas Dynamics* (Wiley, 2002)
6. A.H. Alami, K. Aokal, J. Abed, M. Alhemyari, Low pressure, modular compressed air energy storage (CAES) system for wind energy storage applications. Renew. Energy **106**, 201–211 (2017)
7. A.H. Alami, A.A. Hawili, R. Hassan, M. Al-Hemyari, K. Aokal, Experimental study of carbon dioxide as working fluid in a closed-loop compressed gas energy storage system. Renew. Energy **134**, 603–611 (2019)

Buoyancy Work Energy Storage (BAES) Systems

8.1 Buoyancy Work Energy Storage (BAES) Systems

A promising new energy storage technology that is fit for maritime mechanical storage of off-peak supply of wind farms capitalizes on the work of a buoyancy force applied on a float. The implementation of such systems is facile, especially once appropriate anchoring provisions are integrated in the foundations of wind turbines while they are being built. The large-scale application, where streamlined buoys are coupled with generators of a wind turbine in order to drive it below the sea surface through a hook/pulley assembly when the demand is low and supply is available. The buoys can later be released and allowed to ascend toward the surface when supply requires augmentation by providing the generator with kinetic energy through the cabling system. A depiction of the components and operation of such a system is shown in Fig. 8.1a, b.

The principle of operation of BWES systems is straightforward and is based on the ancient principle of Archimedes. The buoyancy force is a force experienced by a body that is fully or partially immersed in a fluid and is equal to the weight of the displaced fluid that would occupy the same immersed volume. For example, consider the hot air balloon shown in Fig. 8.2.

The flame heats up the air inside the balloon (and keeps it hot). And just like most materials once heated up, the hot air expands (its volume increases) which means that its density will decrease. Since the balloon is immersed in fluid (cold air) which is now denser than the hot air contained within the balloon membrane, there is a net force driving the balloon upward. This net force is the resultant of the buoyancy force, F_b, and the weight of the balloon and can be calculated as

$$F_b = \rho_{\text{hotair}} V_{\text{balloon}} g \qquad (8.1)$$

The density of air at 20 °C is around 1.204 kg/m^3, while it is 0.8986 kg/m^3 at 120 °C (the estimated temperature of air in a hot air balloon). Note that the net force of the balloon

would have to include its weight (neglecting the basket and the operator), and this is calculated as [1]

$$W = \rho_{\text{coldair}} V_{\text{balloon}} g \qquad (8.2)$$

The net force is then $W - F_b = (\rho_{\text{coldair}} V_{\text{balloon}} g) - (\rho_{\text{hotair}} V_{\text{balloon}} g)$ or

$$F_{\text{net}} = (\rho_{\text{coldair}} - \rho_{\text{hotair}}) V_{\text{balloon}} g \qquad (8.3)$$

Equation (8.3) indicates that with the volume of the body being constant, the bigger the density difference is, the bigger the expected net force acting on the body. The investment in a buoyancy-driven system would then be more feasible if the density difference is maximized.

Returning to the BWES system concept shown in Fig. 8.1, the variables that are under the control of the designer are the buoy shape and material, since these are the factors that directly affect its volume and density, respectively. And since the application has already been identified as a maritime one, then the fluid within which the buoy would be immersed is going to be water, which has a nominal density of 998.2 kg/m^3 at 20 °C. A piece of Styrofoam has a density of 50 kg/m^3, which means that if a 1 m^3 sphere of Styrofoam is immersed in water it will experience a net force differential in favor of floating of around 9500 N. This is a major motivation to utilize buoyancy forces and the work resulting from their linear motion (remember that work [J] = force × distance) in energy storage applications.

8.2 Buoyancy Work Theoretical Modeling

A free-body diagram of a generic buoy geometry is shown in Fig. 8.3. The buoy is selected as a streamlined truncated cone to reduce drag forces during operation. The drag force, F_d, will always act opposite to the direction of motion. So, during system charge (buoy pulled downward), the drag will be in the direction of the buoyancy force, while during

A. H. Alami, *Mechanical Energy Storage for Renewable and Sustainable Energy Resources*,
Advances in Science, Technology & Innovation, https://doi.org/10.1007/978-3-030-33788-9_8

Fig. 8.1 A depiction of the operation of BWES system **a** charged and **b** discharged

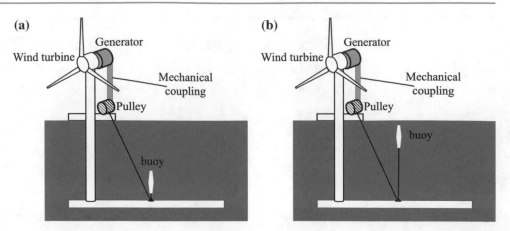

discharge (buoy ascending), the drag will be opposite to it and in direction of the buoy weight, W.

According to the free-body diagram of Fig. 8.2, the dynamic balance of forces acting on each buoy in its direction of motion can be written as

$$F_b - W - F_d = ma \qquad (8.4)$$

where F_b is the buoyancy force, W is the weight of the buoy, and F_d is the drag force, estimated from the drag coefficient. The drag coefficient, C_d, is chosen from values listed in basic fluid mechanics texts. There is no value for a truncated cone; thus, the chosen value will be taken as an average between that of a finite cylinder and a cone (for $\theta \leq 30°$) as

$$C_d = \frac{0.95 + 0.5}{2} = 0.725 \qquad (8.5)$$

8.3   Experimental System Design

To study the expected discharge rate and behavior of the buoyancy system, an experimental rig was constructed that consists of a square $70 \times 70 \times 150$ cm Plexiglas tank filled with water, a buoy made of Styrofoam material and a nylon wire wound around a pulley connected to a 3 W generator through an anchor fixed at the bottom of the water tank, as shown in Fig. 8.4.

The buoys were CNC machined to the truncated cone shape shown in Fig. 8.3. The voltage and current are recorded by virtue of the same NI USB 6009 DAQ system set at 100 Hz for 300 samples (3-second interval). The electrical output (current and voltage) from the generator is recorded by an NI 6009 USB DAQ module at a 500 Hz sampling rate. The current is measured through a connection to a 1-ohm shunt resistor, R (current = voltage/R). The unit is connected to a computer via USB connection and is used to log the data as shown in Fig. 8.5.

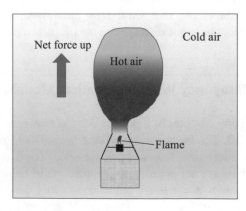

Fig. 8.2 Hot air balloon

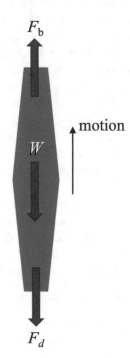

Fig. 8.3 Free-body diagram of a buoy shaped as a truncated cone [3]

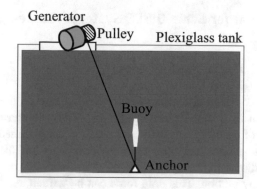

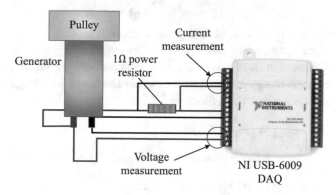

Fig. 8.4 Experimental setup of the BWES system [4]

Fig. 8.5 Electrical wiring diagram of the data acquisition system

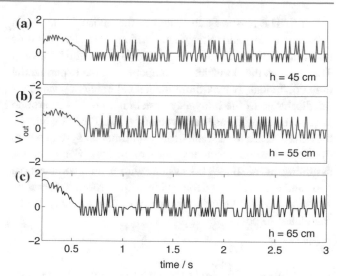

Fig. 8.6 Voltage variation during the discharge cycle at various immersion depths **a** 45.0, **b** 55.0, and **c** 65.0 cm [4]

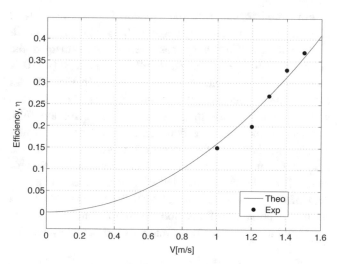

Fig. 8.7 Theoretical versus experimental efficiency of the BWES system [4]

8.3.1 Experimental System Results

The immersion depth of the platform below the surface representing the starting potential energy is varied at 45.0, 55.0, and 65.0 cm, and the generated voltage for each case is recorded. When the buoy is immersed and held under the surface of water at a depth, h, maximum potential energy is reached when the terminal velocity, v, of the buoy near the surface is reached. Kinetic energy can then be released when needed from translational velocities as the buoy system resurfaces. The kinetic energy term for the buoy is given as follows:

$$E = \frac{1}{2} \rho_{\text{buoy}} V_{\text{buoy}} g v^2 \qquad (8.6)$$

Finally, the following relation defines the efficiency of conversion from potential to kinetic:

$$\eta = \frac{\frac{1}{2} \rho_{\text{buoy}} V_{\text{buoy}} g v^2}{h.F_T} \qquad (8.7)$$

where F_T is the total dynamic force applied on the buoy.

For the 3-second discharge time during which buoys are allowed to ascend from various depths, it is noticed that any increase in immersion depth does not have a direct effect on the magnitude and variation of the output voltage. From Fig. 8.6,

one can see the high onset of output voltage for all three cases and then it decreases rapidly as terminal ascension velocity is reached at which point the voltage fluctuates at ± 1.5 V. Also, the output voltage generally starts at a higher value (close to 2 V) for the maximum depth of 65.0 cm indicating a higher quality of the produced voltage. A rectifying circuit could also be used to eliminate the alternating voltage values.

The theoretical and experimental kinetic energies are shown in Fig. 8.7 at which a terminal velocity of 1.5 m/s is reached.

For average velocities of more than 1 m/s, the efficiency of the system approaches 20%. The theoretical efficiency remains higher than the experimental one up to velocities of around 1.4 m/s, at which value the experimental efficiency exceeds the quadratic theoretical projection shown in

Fig. 8.6. The reason is attributed to the inviscid interaction between water and Styrofoam which is a hydrophobic material and would not be wetted by the fluid (water). This reduces the skin friction component of the experimental drag coefficient. A maximum theoretical efficiency value for the discharge of the buoyancy system is found to be around 36% at the terminal ascension velocity of 1.5 m/s. Of course, these results are for a small-scale experimental system where the depths could reach few tens of meters. For example, assuming an ideal BWES system with a unit volume float cube and a 1 m charge depth submersed in room temperature water ($\rho = 998.2$ kg/m^3), the energy storage capacity can be calculated to be

$$E = \Delta \rho g V_{\text{buoy}} \Delta Z$$
$$= \left(998.2 - 50\frac{\text{kg}}{\text{m}^3}\right)\left(1\,\text{m}^3\right)\left(9.81\frac{\text{m}}{\text{s}^2}\right)(1\,\text{m}) = 2.58\,\text{Wh}$$

This energy density is equivalent to that of pumped hydro energy storage.

For another experimental system [1] within a tank of ~ 25 m^3 volume, a buoy height of 0.33 m, a charge depth of 1.67 m, once the buoy has been fully submerged the discharge force remained constantly within a 100–120 N range regardless of charge depth as shown in Fig. 8.8. All measurements recorded at intervals of charge depth were within the margin of measurement of error for the fully submerged buoy. As the buoy emerges beyond the water surface, the discharge force decreases rapidly and reduces to zero at the point where the float is fully floating atop of water. It can be concluded that discharge force is independent of depth for values larger than 0.4 m. This force is a strong function of submersion depth for values less than 0.4 m, which is the scale for many experimental systems.

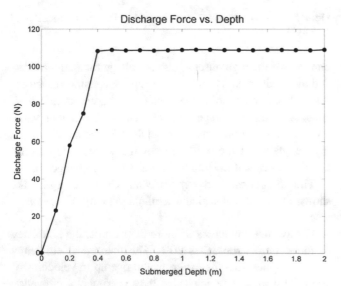

Fig. 8.8 Discharge force versus charge depth [1]

8.4 Larger Scale BWES Systems' Losses [2]

8.4.1 Hydrodynamic losses

When considering hydrodynamic forces acting on the buoy through its motion during charge and discharge phases, the shape of the buoy and its speed must be well defined. In a realistic model, the buoy will perform work on the fluid proportional to the hydrodynamic drag force, F_d, opposing the buoy motion. This drag force can be written as

$$F_{\text{drag}} = \frac{1}{2}\rho A v^2 C_d \tag{8.8}$$

where A is the area of float perpendicular to motion, and C_d is the drag coefficient. Remember that F_d acts against the float motion during both the charge and discharge phases; the total energy loss will be the sum of the losses for charge and discharge over a distance ΔZ:

$$E_{\text{Drag}} = (F_{d1} + F_{d2})\Delta Z$$
$$= \frac{1}{2}\rho A\left(v_{\text{charge}}^2 C_{d1} + v_{\text{discharge}}^2 C_{d2}\right)\Delta Z \tag{8.9}$$

For symmetric buoys, this reduces to

$$E_{\text{Drag}} = \rho A + v^2 C_d \Delta Z \tag{8.10}$$

For the experimental work reported in literature to date, water has been confined within a tank for all experiments reported. Water in such conditions is assumed to be free of inadvertent turbulence, but the actual maritime conditions where wind turbines are located are mainly turbulent, and this would result in inadvertent increase in cable tension for connections with the buoy to the system during both charge and discharge operations. The hydrodynamic drag contributes to the increase and with it the required input torque on the pulley. This would also reduce the output torque during the discharge phase which will consequently decrease the output power.

8.4.2 Acceleration losses

A protocol required to control the resulting torque when releasing the buoy during discharge and the torque required to coil the reel during charging will help mitigate the inertia forces that opposes the change of buoy motion. Energy will be lost at the beginning of both the charge and discharge cycles as more force is required to accelerate the float to the speed required to meet desired power input or output. This energy can be recovered for the charge phase by removing the reel input torque before the achieving the desired charge depth, such that the final meters of charge depth are gained through float inertia. The same principle must be applied at the end of discharge phase with the final meters achieved under increased reel torque.

8.4.3 Mechanical losses

Friction in the reel bearing assembly and pulley will increase the force required during charge and reduce the output force during discharge. This adversely affects the overall efficiency at a rate proportional to the angular speed of the reel and pulley assemblies. Losses can be calculated for a given bearing assembly using well-established equations of bearing design.

8.4.4 Reel and pulley anchorage

The anchorage provisions used for the buoy system have to be installed during the infrastructural works of installing the wind turbine. Since the fixed anchorage of the reel structure to the waterbed may not be feasible in cases of great water depth, a floating platform can be utilized. In case of a floating platform, the effects of rising waters due to wave and tidal motion must also be considered. The anchorage required for the pulley can alternatively be achieved through the use of a large foundation mass to which the pulley is attached, which can be a concrete structure or large rock basket. The required foundation mass will be proportional to ambient fluid velocity, float volume, and design safety factor. For zero ambient fluid, velocity mass, M, required can be expressed as

$$M = \frac{2\left((\rho V g - mg)S_f\right)}{g} \tag{8.11}$$

where M is the foundation mass, and S_f is the safety factor. With the foundation mass correctly designed, the pulley can be deployed from the water surface and left to fall under its own weight to its final position at the waterbed. Torpedo piles, a promising means of offshore anchorage using large torpedo-like anchor piers, could also be used as a cost-effective means of deployment.

8.4.5 Buoy considerations

In order to mitigate drag forces, the buoy should have a streamlined design and be symmetrical about both the horizontal and vertical axes and have a shape that would minimize the drag coefficient. It is known that the hydrostatic forces increase with depth, and thus more stresses are applied on the buoy as it descends further. If the buoy was inflatable, a decrease in volume as the buoy submerges will

increase and thus more losses in energy are incurred. In such cases, the inflatable buoy must possess an internal structure to support its shape against the large external hydrostatic force. A balloon-type float with internal pressure greater than the maximum hydrostatic pressure at max float depth could also be utilized as a constant volume float.

8.4.6 Spatial considerations

For large-scale BWES systems, float volume requirements can be estimated for the ideal case of experimental storage system of a required energy storage capacity. For a unit-radius cylindrical float, a float length of 1.17 m is required for the storage of 1 kWh when deployed in a water depth of 100 m and a marine area of 2.34 m^2 is required.

8.4.7 Electrical losses

When an electric motor and generator are used in conjunction with the reel/pulley system, additional losses will be experienced. The power input or output from the motor unit will be proportional to system voltage and current. The required power level and thus amperage can be calculated for a given float and water depth as

$$i = \frac{(\rho V g - mg)v}{V_{\text{out}}} \tag{8.12}$$

where i is the current in amp, v is the buoy velocity, and V_{out} is the system output voltage. The resistive losses within the motor due to the passing of current can be estimated as

$$E_{\text{electrical}} = \left(\frac{Tv}{V_{\text{out}}}\right)^2 Rt \tag{8.13}$$

where T is the tension in the buoy cable, R is the electrical resistance, and t is the time, which can be written in terms of the distance traveled by the buoy, $\Delta Z = vt$ as

$$E_{\text{electrical}} = \frac{T^2 v}{V_{\text{out}}^2} R\Delta Z \tag{8.14}$$

For equal charge and discharge power levels, and when charge and discharge occur through the same electric motor (i.e., equal resistance of both charge and discharge phases), the total loss can be expressed as twice the value of the one calculated via Eq. (8.14). These losses can be minimized through proper motor selection such that high system voltages are used and electric motor winding resistance is minimized. This does not include inductance losses which require motor specifications for calculation.

8.4.8   Total roundtrip efficiency

With all major system losses accounted for, the total roundtrip efficiency of a BWES system can be expressed as

$$\eta = 1 - \frac{\rho A v^2 C d}{T} - \frac{2TvR}{V_{out}^2} \qquad (8.15)$$

which is applicable for cases of equal charge and discharge speed, and a symmetric float geometry.

References

1. K. Bassett, R. Carriveau, D.S.-K. Ting, Underwater energy storage through application of Archimedes principle. J. Energy Storage **8**, 185–192 (2016)
2. A.H. Alami, Experimental assessment of compressed air energy storage (CAES) system and buoyancy work energy storage (BWES) as cellular wind energy storage options. J. Energy Storage **1**, 38–43 (2015)
3. A. Hai Alami, Analytical and experimental evaluation of energy storage using work of buoyancy force. J. Renew. Sustain. Energy **6** (1), 013137 (2014)
4. A.H. Alami, H. Bilal, Experimental evaluation of a buoyancy driven energy storage device. Adv. Mater. Res. **816–817**, 887–891 (2013)

9.1 Recent Innovations and Applications of Mechanical Energy Storage Technologies

The discussion into mechanical storage technologies throughout this book has entailed technologically simple, yet effective energy storage methods. All technologies share an intuitive implementation philosophy that makes the operation of such techniques be the most cost-effective of other competing storage techniques.

In this chapter, some recent commercial applications are introduced and discussed, which will pave the way for future energy storage-oriented professionals to follow up on, enhance, and hopefully come up with similar novel storage technologies. Also, an honorable mention will be given to two mechanical energy conversion technologies, namely, tidal and wave energy conversion just to complete the discussion. Although the storage element is not obvious in these two technologies, and hence no separate chapters were devoted to them in this book, these technologies deserve our attention. And just like wind energy conversion, which is purely mechanical in nature, the details of construction and utilization of these technologies are left for specialized courses on each respective subject.

9.2 Wave and Tidal Power and Their Storage

Tidal power uses the elevation differential between low and high tides to generate electricity. The gravitational pull of the moon and sun along with the rotation of the earth cause tides. In some places, tides cause water levels near the shore to vary up to 11 m. It is different from the "run-of-river" energy generation in that the supply of the latter is more continuous than the intermittent nature of the former. Although it is outside the scope of this book to study the turbomachines used for energy conversion of water flow from hydraulic potential to kinetic to electrical, it is advantageous to present a brief description of the main ones and

the differences between them. The oldest machine of energy converters due to water flow (or water falling from elevation) is the Pelton wheel, as shown in Fig. 9.1a. It is known as an impulse machine, where a jet of water issuing from a nozzle strikes the blades of the turbine tangentially, causing the wheel to turn and thus generate electricity through a coupled electrical generator.

The another class of machinery is the impulse machines, where the water jet enters the device and leaves in a radial direction (like a water sprinkler where water enters in the z-direction and leaves through a nozzle in the x-y plane, rotating the wheel around the z-axis). The two machines operating on such principle are the Francis and Kaplan turbines, shown in Fig. 9.1b, c. The Kaplan turbine is used for situations where the available water head is low and thus the rotor has a propeller-type design to obtain high angular speeds. The angle of attack of the water jet impinging on the rotor can be optimized by installing guide vanes at the entrance to the rotor region. If the blade pitch angle is fixed, it is called a Nagler turbine. If it can be varied, it is called a Kaplan turbine. The actual efficiencies for the types of turbines described above are shown in Fig. 9.2 as functions of the power level. The design point for these turbines is about 90% of the rated power, so power levels below this point correspond to situations in which the water head is insufficient to provide the design power level.

As for tidal power, the pioneers in this field are EDF, who have been harnessing this energy since 1966 at their plant built across the inlet of the La Rance River in Brittany, France. As one of the very first tidal utilization projects (it is in fact one of just two such plants in the world along with Sihwa in South Korea), it produces around 500 GWh/year with turbines placed in a dam structure across an inlet that serves as a reservoir for two-way operation of the turbines (filling and emptying of the reservoir). The turbine efficiency is reported to be over 90%, but the turbines are only generating electricity during part of the day which is determined by the tidal cycle. Modifications can also be installed to better follow load variations, for example, by using the

A. H. Alami, *Mechanical Energy Storage for Renewable and Sustainable Energy Resources,*
Advances in Science, Technology & Innovation, https://doi.org/10.1007/978-3-030-33788-9_9

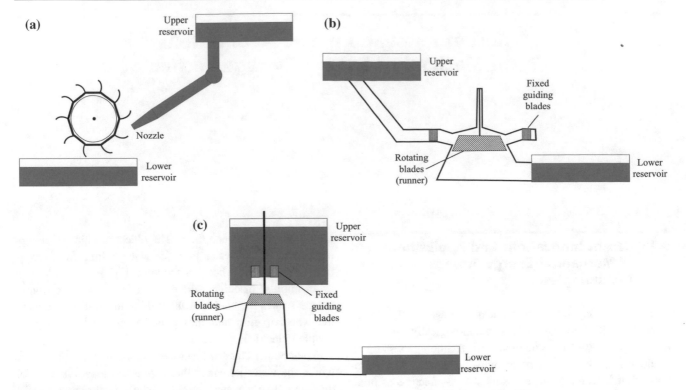

Fig. 9.1 Examples of water-driven turbomachinery **a** Pelton wheel, **b** Francis turbine, and **c** Kaplan turbine [1]

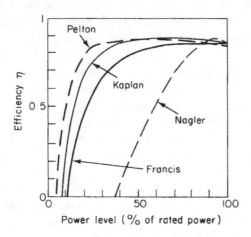

Fig. 9.2 Efficiency of water turbines as a function of power level [2]

turbines to pump water into the reservoir at times when this would not occur as a result of the tidal cycle itself.

Tidal energy can be harnessed by exploiting the differential between low and high tides. The main advantages of using tidal energy to generate electricity are as follows: (i) tides are a perfectly predictable phenomenon (unique for a renewable energy source), (ii) they are inexhaustible and carbon-free, and (iii) they have low environmental impact. The principle involves building a barrage to create an artificial reservoir and so a differential in water levels is created and is capable of driving the turbines and alternators that will

generate electricity. A tidal power plant uses the rising and falling movement of tides to create the level differential needed to produce energy. The barrage is installed across an inlet of an ocean bay or lagoon that forms a tidal basin. Channel gates on the barrage are used to control water levels and flow rates in order to allow the tidal basin to fill on the incoming high tides and then evacuate through a water turbine system coupled with electrical generators on the outgoing ebb tide. A two-way tidal power system generates electricity from both the incoming and outgoing tides.

The La Rance plant in France has an installed capacity of 240 MW distributed between 24 bulb-type turbine generators, each with a capacity of 10 MW. For almost 50 years, it has been producing around 500 GWh/year, equivalent to the consumption of a city, the size of Rennes, France. The barge-like design of the station is shown in Fig. 9.3.

The La Rance plant operates according to two different modes:

1. Ebb generation: The six valves of the barge are opened until high tide is reached to fill the reservoir. As the tide recedes, the differential created between the stored water and the sea is sufficient to drive the turbines and alternators to generate electricity
2. Flood and ebb generation: During spring tides, electricity is also produced on the incoming as well as on the outgoing tides. The differential between the low tide and the

Fig. 9.3 Tidal installation of La Rance River in Brittany [3, 4]

level upstream from the barrage is sufficient to drive the bulb-type turbines and generate electricity on the flood tide. The turbines and alternators were specifically designed to be able to operate in both directions.

As an energy storage option, La Rance plant turbines can also help pump water. The associated generator can operate in reverse mode and pumps seawater in order to further raise the level behind the barrage before slack water is reached (the short period when there is no movement either way in the tidal stream). This enables production to start up sooner and so maximize output.

Another storage option devised in conjunction with tidal power is proposed by TM Power, who recently won a competitive tender to supply an integrated hydrogen system for use at the European Marine Energy Centre (EMEC) tidal test site on Eday, Orkney, Scotland. The principal component of the system is a 0.5 MW polymer electrolyte membrane (PEM) electrolyzer with integrated compression and up to 500 kg of storage. The 0.5 MW electrolyzer will be used to absorb excess power generated by the tidal turbines testing at EMEC [5]. The hydrogen gas generated will be compressed and stored, with some of the gas being used in (an optional) hydrogen fuel cell to provide backup power to critical EMEC systems. The remainder of the hydrogen gas will be used off-site by a further project being developed separately which plans to absorb output of a local community wind turbine operated by Eday Renewable Energy Ltd. Hydrogen generation capacity is estimated to be up to 220 kg/24 h, and the device is self-pressurizing up to 20 bar with hydrogen purity satisfying ISO 14687.

9.3   Adiabatic Compressed-Air Energy Storage for Electricity Supply (ADELE)

ADELE is a large-scale CAES storage developed and operated by the German RWE Power company. The project is classified under AA-CAES systems and has the objective of operating at efficiencies of around 70%. There will be no combustion processes involved, and thus no burning of fossil fuels will permit a CO_2-neutral production of peak load electricity from renewable energy. A commercial plant is expected to store 1,000 MWh of electrical energy and feed some 300 MWel (megawatt electric) into the grid for several hours. The details of ADELE components are shown in Fig. 9.4.

General Electric (GE) is developing the air compressor and air turbine, two of ADELE core components. Driven by an electric motor, ambient air is at the intake of the compressor which has a pressure ratio of around 100. The air is then fed into the heat storage device as hot compressed air at around 600 °C. The heat is stored in the TES silos shown in Fig. 9.5 for later use during the discharge to preheat the expanding air leaving the cavern storage. This compressed air will flow into the air turbine and force it to rotate and drive the coupled electrical generator. The compressor and turbine are expected to have efficiencies of at least 70%.

The heat storage silos are containers with heights reaching up to 40 m high, lined with stones or ceramic molded bricks through which the hot air flows. Ed. Züblin AG are in charge of developing the heat exchangers and TES systems for ADELE, where they have to make sure that it is capable

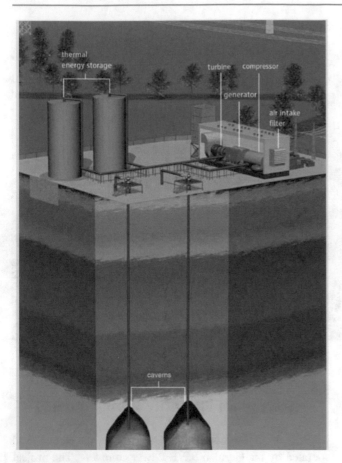

Fig. 9.4 Depiction of the operation of ADELE [6]

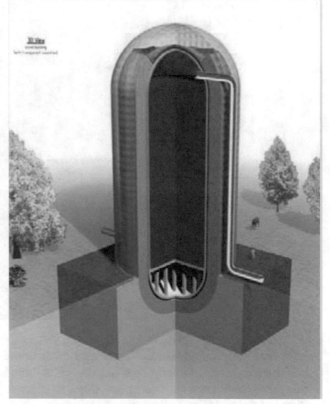

Fig. 9.5 Energy Vault (potential energy storage using gravity)

of handling the high pressures and cyclic temperature, which may lead to premature thermal fatigue failures.

This system operates on a principle similar to pumped hydro storage. The water here is replaced with cylindrical blocks, and a crane takes the place of water pumps (see Fig. 9.6a). Whenever energy is abundant, especially if the source was a renewable one (mainly wind energy as shown in Fig. 9.6b), the crane stacks the blocks at various heights using a motor/pulley system. Once energy demand rises, energy is called from storage by lowering the blocks with the crane, where the motor acts as a generator, and the potential energy is transformed into kinetic and eventually into electrical energy. The system operation is shown in Fig. 9.7, where (a) is the fully charged (blocks stacked high) and (d) is the fully depleted storage system.

The bricks are made from a sustainable composite material, and the total weight of the bricks reaches up to 35 tons. The system is modular and promises flexibility with plant capability ranges of 20–35–80 MWh storage capacity and a 4–8 MW of continuous power discharge for 8–16 h. The system is very simple, will not degrade over the life of the project it serves, and has competitive performance with roundtrip efficiency between 80 and 90%.

9.4 Cryogel Thermal Storage System

Cryogel is a thermal energy storage (TES) technology, which is registered under Airclima Research in Paris, France. The system operates as a thermal storage for cooling applications, which can involve slightly different provisions and calculations than thermal storage for heating applications. The technique utilizes small ice capsules that has an encapsulated cryogenic material, in conjunction with the ice spray system. During times of low demand, chillers that supply cold water are not shut off, but rather are rerouted to cool down the balls down to (or near) the freezing temperature of the cryogenic material. This stage can take place all long (when demand is minimal), and hence the chiller can operate at (or lower than) standard base capacity. During times of high demands, the chiller supplies users, but the thermal storage system can also respond by allowing the spray system to pass water through the ice capsules and thus assisting with the cooling load. This results in a more homogenous loading of the chiller in face of the variable demand. The system components are shown in Fig. 9.8.

The ice capsule dimensions along with some properties are shown in Table 9.1, with a picture of one shown in Fig. 9.9.

Fig. 9.6 **a** Energy Vault system, **b** a depiction of the system integration with wind energy [7]

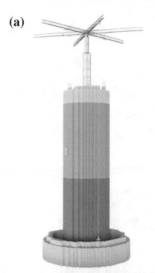

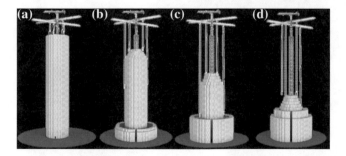

Fig. 9.7 Energy Vault system at various stages of charge: **a** is fully charged and **d** is fully discharged

An installed capacity of 500 kW of ice thermal storage is attached to a 1000 kW chiller, with a reported storage capacity of 4000 kWh and storage volume of 90 m³ of ice

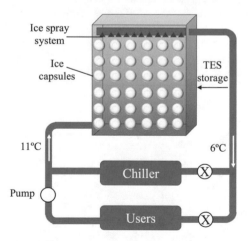

Fig. 9.8 Elements of cryogel ice spray system

Table 9.1 Properties of ice capsules [8]

Property	Value
Diameter (mm)	137
Total weight (g/piece)	1085
Weight of water (g/piece)	960
Number of pieces (kWh)	11.24
Energy density at 0 °C (kWh/m^3)	46.30
Exchange surface (m^2/kWh)	0.8

Fig. 9.9 The shape of an ice capsule [8]

capsules. The advantages of using a spray system are reduced quantities of brine (water–glycol mixture) required to pass through the refrigeration cycle; the tanks can be made of any material (concrete, metal, etc.) and assume any shape according to the available space. There is no hydrostatic pressure due to prolonged water storage in the tanks. And the system promises over 46 kWh/m^3 of storage density.

References

1. B. Sørensen, *Renewable Energy Conversion, Transmission, and Storage*. Elsevier/Academic Press (2007)
2. G. Fabritz, Wasserkraftmaschinen, in *Hütte Maschinenbau*, vol. IIA (Wilhelm Ernst and Sohn., Berlin, 1954)
3. Tidal Power| EDF France. https://www.edf.fr/en/the-edf-group/industrial-provider/renewable-energies/marine-energy/tidal-power. Accessed 22 Aug 2019
4. Tidal power—U.S. Energy Information Administration (EIA). https://www.eia.gov/energyexplained/hydropower/tidal-power.php. Accessed 22 Aug 2019
5. Tidal Energy Storage| ITM Power. http://www.itm-power.com/project/tidal-energy-storage. Accessed 22 Aug 2019
6. Adele-Adiabatic Compressed-Air Energy Storage For Electricity Supply RWE Power. https://www.rwe.com/web/cms/mediablob/en/391748/data/364260/1/rwe-power-ag/innovations/Brochure-ADELE.pdf. Accessed 15 Sept 2019
7. Energy Vault—Ground-breaking energy storage technology enabling a planet powered by renewable resources. https://energyvault.com/. Accessed 22 Aug 2019
8. Ice Thermal Storage—CRYOGEL. http://www.airclima-research.com/. Accessed 22 Aug 2019

Printed in the United States
By Bookmasters